T0135494

Communication and Localization

in UWB Sensor Networks

A Synergetic Approach

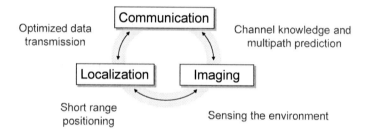

Optimized data
transmission

Channel knowledge and
multipath prediction

Short range
positioning

Sensing the environment

Heinrich Lücken

Reihe Series in Wireless Communications
herausgegeben von:
Prof. Dr. Armin Wittneben
Eidgenössische Technische Hochschule
Institut für Kommunikationstechnik
Sternwartstr. 7
CH-8092 Zürich

E-Mail: wittneben@nari.ee.ethz.ch
Url: http://www.nari.ee.ethz.ch/

ISBN 978-3-8325-3332-8
ISSN 1611-2970

Logos Verlag Berlin GmbH
Comeniushof, Gubener Str. 47,
10243 Berlin
Tel.: +49 030 42 85 10 90
Fax: +49 030 42 85 10 92
INTERNET: http://www.logos-verlag.de

Diss. ETH No. 20776

Communication and Localization in UWB Sensor Networks

A Synergetic Approach

A dissertation submitted to the
ETH ZURICH

for the degree of
Doctor of Sciences

presented by
HEINRICH LÜCKEN
Dipl.-Ing., RWTH Aachen
born May 20, 1981
citizen of Germany

accepted on the recommendation of
Prof. Dr. Armin Wittneben, examiner
Prof. Dr. Lutz Lampe, co-examiner

2012

Abstract

In this thesis, a novel sensor network paradigm is proposed and studied, inspired by the fusion of wireless communication, localization and imaging. Wireless sensor networks will open a fascinating world of ubiquitous and seamless connectivity not only between individuals but also between devices and objects in our daily life. The key to this vision is a universal low-power, low-complexity and low-cost transceiver unit that provides scalable data communication as well as location and environmental information. Ultra-Wideband (UWB) technology with its rich design space can meet the challenging requirements of future wireless sensor networks. This is the consequence of a paradigm shift compared to narrowband communication: due to the huge bandwidth available, we can trade off bandwidth efficiency against other figures of merit. The major design criterion is not data rate anymore, but rather power consumption and hardware complexity. Within the group of hardware-aware system designs, UWB impulse radio with energy detection receivers are of particular relevance and well known for their efficient implementation. The contribution of this thesis is the comprehensive study of sensor networks with generalized energy detection receivers, where we focus on innovative and efficient approaches for communication and localization and their synergy.

The first part of this thesis develops a framework for location-aware optimization of data transmission with generalized energy detection receivers. This framework is based on a Signal-to-Interference-plus-Noise-Ratio expression. It covers receiver as well as transmitter optimization, where narrowband interference suppression is also taken into account. Conventional approaches attempt to adapt the transceiver directly to the channel state. They require the knowledge of the channel state over the full transmission bandwidth. Due to the huge bandwidth of UWB, the estimation and dissemination of channel state information requires high complexity and is very expensive and power hungry. To circumvent this problem, we propose adjusting the transceiver to the node position. This is done by modeling the channel impulse response as a random process with location dependent parameters, which can be estimated in an off-line training phase. The data transmission is then optimized based on the position of the node – a more accessible type of information in UWB networks that may already be available. In the next step, we extend the optimization to multiuser transmission. This leads to an increase in the sum data rate, while maintaining the low complexity of the nodes. We conclude from performance evaluations that location information can improve the performance of low complexity and low power UWB communication.

The second part of this thesis is dedicated to localization. We focus on the estimation of the travel time of the radio signal, which is related to the distance and, thus, to the location of the node. Many existing approaches for time of arrival (ToA) estimation of UWB signals require high speed sampling of at least twice the bandwidth of the received signal. This leads to high power consumption and high hardware complexity. We propose to perform ToA estimation at the output of an energy detection receiver. This allows the sampling rates to be much lower keeping the complexity and the power consumption low. In order to achieve high performance, we derive the maximum likelihood timing estimator for the generalized energy detection receiver. We again model the channel impulse response as a random process and show that location-aware a-priori knowledge of its distribution can increase the performance significantly. Approximations of the maximum likelihood estimator lead to a family of low-complexity timing estimators that trade lower estimation accuracy for lower computing time. Additionally, we analyze spectral timing estimation at the energy detector output. We derive the accuracy in a multipath channel analytically, which provides an insight into the fundamental performance scaling of the estimation error in multipath.

So far, we show that location-aware a-priori channel knowledge is beneficial for communication as well as position estimation. However, this requires a database that maps positions to channel characteristics. The training phase to acquire this database can be difficult to implement, time-consuming and may require many training samples. The solution to this problem is imaging. In the third part of this thesis, we present radar imaging based channel prediction. Using this method, the channel characteristics can be obtained for every position in a stationary environment from just a few training samples. An extensive measurement campaign proves the practicality of the presented approach. Finally, we draw conclusions and give an outlook on future research in UWB communication, localization and imaging.

Kurzfassung

Inspiriert durch die Verschmelzung der drahtlosen Kommunikation, Ortung und Bildgebung wird in dieser Arbeit ein neuartiges Sensornetzwerk-Konzept vorgeschlagen und untersucht. Zukünftige drahtlose Sensornetzwerke werden eine faszinierende Welt eröffnen, die allgegenwärtige und nahtlose Vernetzung ermöglicht und das nicht nur zwischen einzelnen Personen, sondern auch zwischen den Geräten und Gegenständen in unserem täglichen Leben. Der Wegbereiter zu dieser Vision ist ein universeller Funk-Baustein, der sich durch geringen Stromverbrauch, geringe Komplexität und geringe Kosten auszeichnet und sowohl skalierbare Datenübertragung erlaubt als auch Positions- und Umgebungsinformationen zur Verfügung stellt. Ultra-Wideband (UWB) Technik kann durch den vielfältigen Designraum den hohen Anforderungen der zukünftigen drahtlosen Sensornetzwerke gerecht werden. Dies ist insbesondere die Folge eines Paradigmenwechsels im Vergleich zur herkömmlichen schmalbandigen Kommunikation: Durch die große frei verfügbare Bandbreite können wir Bandbreiteneffizienz gegen andere Kenngrößen abwägen. Somit sind die wichtigsten Entwurfsziele nicht mehr die Datenrate und Reichweite, sondern Stromverbrauch und Hardware-Komplexität. Innerhalb der hardwareeffizienten Systementwürfe ist UWB Impulsfunk mit Energy-Detection-Empfängern von besonderer Bedeutung und für eine effiziente Implementierung bekannt. Der Beitrag dieser Arbeit ist die umfassende Analyse von Sensornetzwerken, die auf verallgemeinerten Energy-Detection-Empfängern basieren. Dabei wird der Fokus auf innovative und effiziente Methoden zur Kommunikation und Ortung sowie deren Synergie gerichtet.

Im ersten Teil der Arbeit wird ein Grundgerüst entwickelt, das die ortsabhängige Optimierung der Datenübertragung mit verallgemeinerten Energy-Detection-Empfängern ermöglicht. Dieses Framework basiert auf einem Störabstandskriterium und umfasst sowohl Interferenzunterdrückung als auch Sender- und Empfängeroptimierung. Herkömmliche Ansätze versuchen, die Sender und Empfänger unmittelbar an den Übertragungskanal anzupassen. Dazu muss der Übertragungskanal über die komplette Bandbreite bekannt sein. Durch die hohe Bandbreite von UWB wird die Schätzung und Verteilung der Kanalinformation sehr komplex, teuer und verbraucht viel Strom. Um dieses Problem zu umgehen, schlagen wir vor, die Übertragung basierend auf der Position des Funkknotens anzupassen. Dazu modellieren wir die Kanalimpulsantwort als Zufallsprozess mit ortabhängigen Parametern, welche in einer offline Trainingsphase geschätzt werden können. Die Datenübertragung wird dann anhand der Knotenposition optimiert – einer leicht zugänglichen Information in UWB Netzwerken, die ohnehin schon verfügbar sein mag. Im nächsten Schritt betrachten wir die Optimierung für Mehrbenutzer-Systeme. Dadurch kann die Summendatenrate erhöht werden unter Beibehaltung der geringen Komplexität der Empfangsknoten. Anhand von Simulationsergebnissen wird gezeigt, dass die UWB Kommunikation mit der Positionsinformation optimiert werden kann, um Stromverbrauch und Komplexität gering zu halten.

Der zweite Teil dieser Arbeit widmet sich der Ortung von Funkknoten. Insbesondere untersuchen wir das Schätzproblem der Signallaufzeit, das in Bezug zur Entfernung und dadurch zur Position des Funkknotens steht. Viele bestehende Ansätze zur Time-of-Arrival (ToA) Schätzung erfordern sehr schnelles Abtasten mit mindestens doppelter Bandbreite des Empfangssignals. Dies führt zu einem hohen Stromverbrauch und einer hohen Hardware-Komplexität des Empfängers. Unser Vorschlag ist es, die ToA Schätzung am Ausgang des Energy-Detection-Empfängers durchzuführen. Dies ermöglicht wesentlich geringere Abtastraten und hält damit die Komplexität und den Stromverbrauch niedrig. Um dennoch eine hohe Genauigkeit zu erreichen, leiten wir den Maximum-Likelihood Timing-Schätzer für den generalisierten Energy-Detection-Empfänger her. Dazu modellieren wir die Kanalimpulsantwort wieder als Zufallsprozess und zeigen, dass Kenntnis über die ortsabhängige Wahrscheinlichkeitsverteilung die Schätzgenauigkeit wesentlich verbessern kann. Näherungen des Maximum-Likelihood Schätzers führen zu einer Familie von Timing-Schätzern mit geringerer Komplexität, die einen Trade-Off zwischen Genauigkeit und Rechenaufwand ermöglichen. Zusätzlich analysieren wir die spektrale Timing-Schätzung am Energy-Detector-Ausgang. Durch die Herleitung eines expliziten Ausdrucks für die Genauigkeit wird ein Einblick zum fundamentalen Verhalten des Schätzfehlers in Mehrwegeausbreitungskanälen gegeben.

Soweit wird gezeigt, dass ortsabhängiges Kanalwissen sehr nützlich ist für die Kommunikation und Ortung. Allerdings wird dazu eine Datenbank benötigt, die Positionen auf Kanaleigenschaften abbildet. Die Trainingsphase, um diese Datenbank zu erstellen, kann möglicherweise kompliziert und zeitaufwändig sein und eine große Anzahl an Stichproben benötigen. Einen Lösungsansatz zu diesem Problem finden wir in der Bildgebung: Im dritten Teil dieser Arbeit präsentieren wir ein Radar-basiertes Verfahren zur Kanalvorhersage. Mit dieser Methode können die Kanaleigenschaften für beliebige Positionen durch Kenntnis einiger weniger Stichproben abgeleitet werden. Eine umfangreiche Messkampagne belegt die Praxistauglichkeit des vorgestellten Ansatzes. Schließlich ziehen wir Schlussfolgerungen und geben einen Ausblick auf die zukünftige Forschung zur UWB Kommunikation, Lokalisierung und Bildgebung.

Contents

Chapter 1

Introduction

1.1 Motivation

Sensor networks have a wide area of applications, which make our daily lives more convenient, more secure and more efficient. Examples for sensor networks range from industry to healthcare to entertainment technology. In industry, sensor networks are necessary in all areas of manufacturing and production. Modern production technology is based on automation and automation requires monitoring and feedback from sensors. More sensor data and information processing increases the productivity and quality of manufacturing. Better monitoring enables more accurate control of machines and engines, which leads to increase in efficiency and to decrease in emissions, pollution and resource consumption. In healthcare, sensor networks are necessary for monitoring vital signs such as pulse rate, blood pressure and respiratory rate. This is essential in hospitals in order to monitor critical patients and to help to save lives. In the future, sensor networks can be included in clothes in order to detect heart attacks or strokes. Additionally, the sensor data can be used to assist in rehabilitation and physiotherapy. For the elderly, sensor networks can check vital information and they can be used for fall detection. For entertainment, sensor networks create new forms of human-computer interaction. Sensor networks attached to the human body can be used to analyze athlete performance or to detect an injury or accident. All these exciting applications of sensor networks require a flexible and reliable communication platform. The sensor data needs to be measured and processed but also to be communicated. A successful deployment of sensor networks requires that the communication is low power, low complexity, and low cost.

Today's sensor networks are either based on wired links or narrowband radio communication. The installation of wires is impractical in many cases due to moving parts, harsh

1

environments or sealed areas. Furthermore, wired data transmission can be error-prone and expensive. The installation of wiring harnesses is complicated and requires a high expenditure of time and work. Cables are expensive and they can break easily. This leads to severe failures, which are complex to repair and thus involves very high maintenance costs. Generally, wireless systems have the potential to solve these problems. However, the state-of-the-art of narrowband communication has two major drawbacks. First, adequate narrowband transceivers are power hungry, complex, and may require multiple antennas and thus a lot of space. Second, conventional narrowband communication suffers from fading of the radio signals. For example in harsh propagation environments, which are typical for the aforementioned sensor networks, the multipath propagation can lead to bad channel conditions for narrowband systems. This is a problem in particular for safety related systems, which need to ensure a low probability of an outage or transmission error.

Ultra-wideband (UWB) communication can meet the stringent requirements of sensor networks. The reason for this is that UWB systems can use up to 7.5 GHz available and admission free radio spectrum. Conventional narrowband systems suffer because they are restricted to a small part of the radio spectrum. This makes radio spectrum to a valuable and expensive resource. The situation is different for UWB and the large bandwidth in the spectrum from 3.1 - 10.6 GHz allows for innovative systems designs. Therefore, UWB communication is one of the most promising technologies for future wireless sensor networks. In addition, UWB technology has two other important features for sensor networks. First, the high bandwidth enables position estimation of sensor nodes with centimeter accuracy. UWB is generally considered the method of choice for facilitating high definition positioning. Second, UWB radar imaging provides useful information about the propagation environment of the sensor network. A high spatial resolution can be achieved because single multipath components are resolvable from the receive signal. Both high definition localization and imaging are a unique feature of UWB and a consequence of the high bandwidth.

In this thesis, we propose algorithms for communication, localization and imaging for short-range UWB sensor networks. The focus is on low complexity, low cost and low power system designs and the synergy of communication and localization.

1.2 Ultra-Wideband Technology

A comprehensive overview on UWB technology is given in [1–3]. In this section, we present a short review and recall the main advantages of UWB for wireless sensor networks.

The history of UWB communication dates back to the very beginning of radio communication. When Guglielmo Marconi[1] invented the radio transmission, he used a spark generator to transmit short pulses and applied a coherer as receiver [4, 5]. Whereas we would consider this today as UWB communication, the early design choices were rather a result of limited hardware and technology. The further evolution of wireless communication was affected by carrier modulated narrowband communication, which may have originated from multiple access by frequency division. The renaissance of UWB communication can be traced back to the late 1990s, when impulse radio was proposed for short-range communication in dense multipath environments [6]. UWB impulse radio uses short pulses with the signal energy spread over a large spectrum. In an operating regime, where conventional narrowband signaling comes to its limitations, UWB was now being proposed as a secondary spectrum user.

Generally, UWB communication is intended to be below the noise floor of narrowband signaling to not interfere with conventional systems [7]. The key advantage of UWB communication is its frequency diversity, which makes it suitable for harsh multipath environments. In contrast to narrowband signaling, UWB does not suffer from multipath fading and shows high robustness even in harsh propagation environments [8]. Just like conventional spread spectrum systems, impulse radio uses a much larger signaling bandwidth than data rate. However, in impulse radio, the spreading is obtained by low-duty cycle signaling. Thus, the transmit signal is sparse in time domain and multiple access can be provided, e.g. by time hopping [9, 10].

Much attention on UWB communication was triggered in February 2002, when the FCC adopted license-free UWB operation in the United States of America [11]. Numerous system architectures and signaling schemes were proposed and industry consortia tried to establish different standards. Most of these approaches can be classified as either high or low data rate systems. High data rate systems were mainly driven by the consumer electronics industry as cable replacement for external devices such as hard disks, video displays or hand-held computers. The most popular of these is probably the ECMA-368 standard [12], which defines the physical layer of Wireless USB Specification 1.1 [13]. It supports data rates of up to 1 Gbps and is based on conventional multiband orthogonal frequency-division multiplexing (OFDM). For low data rate systems, a common example is the IEEE 802.15.4a standard, which defines a UWB impulse radio physical layer for wireless personal area networks (WPANs) [14]. This standard supports data rates of

---

[1]Guglielmo Marconi (1874–1937). Italian inventor, Nobel Prize in Physics (1909) for the development of wireless telegraphy.

0.11 – 27.2 Mbps and targets short-range communication for low-power and low-complexity devices [15, 16]. Many of the approaches that are presented in this thesis could also be applied to systems based on IEEE 802.15.4a. However, to keep the results as generic as possible, we do not limit the considered system to a specific standard.

Localization based on UWB technology is well understood [17–20] and is generally considered the method of choice facilitating high definition positioning for short range wireless networks. In principle, UWB positioning can be classified into geometric approaches and fingerprinting based systems. Geometric approaches include positioning based on time-of-arrival (ToA) or angle-of-arrival (AoA) determination. The position estimation of a mobile agent node relies on fixed anchor nodes with known positions. For the case of ToA position estimation, the ranges between an agent and several anchors are measured [21]. The individual ranges can then be used to determine the agent position by multilateration [22]. Fingerprinting based systems use location specific features of the propagation channel to determine the agent's position. Usually, a set of measured data with known positions is used as training data. The choice of the fingerprint depends on the training data and the hardware, e.g. system complexity. Promising approaches estimate or classify the agent's position based on extracted signal metrics such as signal strength (RSSI) [23] or the shape of the channel impulse response [24, 25].

The state-of-the-art in UWB imaging can be divided into two main approaches: tomographic imaging and radar based systems [26]. Tomography aims to see inside an object, where the object is typically surrounded by a large number of sensors. Backprojection algorithms are used to reconstruct the internal properties. Prominent applications of this technique include microwave breast cancer detection [27], which is much less harmful than the conventional X-ray based checkup. The other approaches of UWB imaging are radar based systems, which aim to retrieve the shape and location of a distant object [28]. The radar image is usually obtained from an antenna array or from moving antennas, i.e. synthetic aperture radar (SAR). The reconstruction problem was tackled by seismic engineering and geophysics in the early 1980s, where migration of seismic data was introduced to locate in-ground resources [29]. Recently, UWB imaging was proposed for through-wall imaging for disaster recovery, e.g. to find trapped humans or cavities after a building collapse [30].

Each of the fields of UWB communication, localization as well as imaging is advanced in its research and many technologies and products are ready for the market. However, the combination and synergy of this trio has great potential. Surprisingly, so far this has not been considered in literature, even though for higher layers many approaches of combined communication and localization have been proposed. Most prominent among these are location-aware services at the application layer, which dynamically provide the user with travel information, shopping, entertainment or event information [31]. Furthermore, location information is also intensively used to improve routing for sensor networks [32] or security protocols [33]. A more abstract example for the synergy of localization and communication is cooperative localization in wireless sensor networks [34, 35] or position estimation based on a connectivity metric of nodes [36]. However, in this work we consider the physical layer signal processing and are interested in optimized transmission schemes, ranging or positioning algorithms and the specific characteristics of the UWB propagation channel.

Extensive research on UWB communication and localization has been performed at the Wireless Communication Group, ETH Zurich. In [37], signaling schemes and low complexity receivers for UWB body area networks are proposed and studied. The contribution [38] introduces low duty cycle signaling for optimized and FCC-compliant UWB communication and presents a low power energy detection receiver. In [39], position estimation by fingerprinting of multipath components is proposed and analyzed. Ref. [37–39] serve as starting points for the results presented in this thesis.

1.3 Contributions

The results presented in this thesis are concerned with communication, localization and imaging for UWB sensor networks, see Fig. 1.1. The concepts are introduced in a generic way and are applicable for different systems and setups. However, in the course of this dissertation, we consider a specific network with many low complexity sensor nodes, which communicate over a short range with stationary cluster heads. The sensor nodes are equipped with generalized energy detection receivers, due to their stringent complexity requirements. For this setup, we aim to answer the following questions:

1. How can communication benefit from a priori channel state information?

2. How can localization benefit from a priori channel state information?

3. How can we obtain a priori channel state information?

The first question inspires a family of optimization algorithms for the transmitter and receiver, which make the data transmission more efficient and reliable. For localization, the a priori knowledge improves the ToA estimation and reduces the position estimation bias and error. The third question leads to one of the key concepts of this work: We establish the mapping between the node location and the channel state information (CSI). In UWB systems, positions or regions can be related to certain multipath characteristics. We introduce

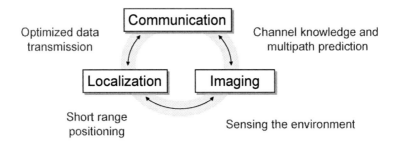

Fig. 1.1: Synergy of UWB communication, localization, and imaging

regional channel knowledge as a vicinity based a priori CSI for the sensor nodes. This is an innovative and efficient approach since channel estimation becomes expensive with increasing bandwidth, while location information is easily accessible. However, this approach requires a database with channel characteristics for different locations. For a stationary environment (as typical for indoor scenarios), this has to be built up in a calibration phase from training data. To keep the number of training samples low, we propose to use UWB imaging to predict channel impulse response for arbitrary positions. Throughout this thesis, we attach importance to the usefulness and practicability of the presented approaches. Furthermore, all approaches are evaluated based on extensive measurements to quantify the performance gains over reference schemes in a realistic environment.

The thesis is structured as follows. The first part of this thesis (Chapter 3 and 4) is concerned with communication, the second part (Chapter 5 and 6) with localization, and the third part (Chapter 7 and 8) with imaging. The results of the thesis have been published in [40–49]. In summary, we make the following contributions:

In Chapter 2, we discuss the advantages and the motivation of UWB for low complexity systems and introduce the system model, which is considered throughout this thesis. We give a short overview on non-coherent UWB systems and elaborate on alternative system designs. The generalized energy detection receiver with binary pulse position modulation (BPPM) is identified as method of choice due to the following reasons: efficient analog implementation, low rate sampling, robustness to channel variations.

In Chapter 3, we derive the signal-to-interference-plus-noise-ratio (SINR) of the generalized energy detection receiver as a function of the transmit pulse and the post detection filter. This enables us to develop powerful optimization algorithms. We include narrowband interference to account for the secondary spectrum usage of UWB communication. The optimization algorithms are derived for both full CSI and regional CSI. A performance evaluation based on measured channel impulse responses shows the gains of the presented transmission schemes compared to conventional energy detection. Parts of this chapter have been published in [40, 46].

In the next step, we extend the optimization to multiuser transmission, which is presented in Chapter 4 and [44]. We focus on the downlink of the sensor network, i.e. data transmission from the cluster head to the low complexity nodes. When considering low complexity receivers, the data rate is limited due to inter-symbol interference from multipath. Conventional approaches for inter-symbol interference mitigation require expensive post processing and thus increase the power consumption and costs of the sensor nodes. To circumvent this problem, we propose to transmit several data streams to different nodes simultaneously. Thus, a higher throughput can be achieved, while maintaining low complexity of the sensor nodes. We derive a precoding scheme for multiuser transmission, which is based on the maximization of the minimum SINR over the nodes. Again we consider both full CSI and regional CSI. Performance evaluation shows small loss compared to time division multiple access (TDMA) for the case of full CSI, even though the data rate is multiplied. The multiuser transmission works even with statistical CSI with a certain performance loss that depends on the characteristics of the region.

In Chapter 5, we derive the maximum likelihood timing estimate for generalized energy detection receivers with statistical CSI. To solve the maximum likelihood estimation problem, we provide a closed form expression of the probability density function (PDF) of the generalized energy detector output. Additionally, we propose a family of low complexity timing estimators that are systematically derived from approximations of this PDF. Timing estimation is essential for both communication and localization. Whereas maximum likelihood timing estimation is well known for linear receivers, it is difficult for non-coherent receivers. For communication, the timing estimation is used for symbol synchronization, i.e. to find the best sampling time for data detection. For localization based on ranging, timing estimation is one of the most crucial components and major sources of error. Whereas coherent UWB receivers require analog-to-digital conversion of up to 15 GHz sampling rate, the sampling at the output of the energy detection receiver can be much lower. By taking the a priori CSI into account, the timing estimation can be more accurate than the sampling time of the receiver. Parts of this chapter have been published in [41, 42].

To reduce complexity, we introduce UWB timing estimation, which can be performed completely in an analog fashion and does not require high rate analog-to-digital conversion.

This is presented in Chapter 6 and [43]. The approach does not require any CSI and is related to spectral timing estimation. We provide an analytical expression for the estimation bias and error. This gives insight into the performance scaling of ToA estimation in multipath channels with respect to the transmission bandwidth. It can be shown that spectral timing estimation can benefit from the high bandwidth and performs superior to narrowband ranging even under the strongest of complexity constraints.

In Chapter 7, we present radar imaging based multipath delay prediction to realize efficiently the mapping from the position to the channel state [45]. We propose a three step approach: First, the radar image of the environment is created with training data. Second, the channel responses are synthesized for arbitrary position. Finally, the dominant path delays are extracted to suppress out-of-band noise. The extracted dominant multipaths are of great value for position estimation and data transmission.

Chapter 8 is dedicated to measurements and experimental results. Namely, we describe two advanced channel measurement systems, where one is designed for high precision and the other for high-speed/low-cost channel measurements. Furthermore, we present an extensive measurement campaign in an office environment. The measured channels enable us to demonstrate and verify imaging based multipath delay prediction. Selected performance results on localization and communication show the benefit of the synergy of communication, localization, and imaging.

Finally, in Chapter 9 we draw conclusions from presented results. We also discuss UWB's outlook and address promising applications of UWB communication, localization and imaging.

Chapter 2

Low Power, Low Complexity, Low Cost - High Bandwidth?

In this chapter, we discuss the requirements of sensor networks and motivate the use of high bandwidth for low data rate systems. Furthermore, we give a brief introduction to non-coherent UWB systems and present the system model which is considered throughout this thesis.

A prominent application and motivation of the results presented in this thesis are wireless sensor networks [50]. Traditionally, sensor networks are often defined as distributed systems to monitor physical or environmental conditions, e.g. the temperature distribution in a building. The communication takes then place between sensor nodes and central entities, which collect the measurement data. However, in the context of this thesis, we want to avoid the limitation to a specific application and would like to consider sensor networks in a much broader sense. In order to define this more precisely, we classify the applications of wireless communication into human-to-human, human-to-machine or machine-to-machine communication. On the consumer market, the early stages of wireless technology were mainly driven by human-to-human communication, e.g. phone services in mobile radio networks. When looking at the last decades, we can observe that the innovations were focusing rather on data services, i.e. human-to-machine communication. An example for this class is mobile broadband Internet or WLAN.

Within this framework, we will consider sensor networks from the class of machine-to-machine communication with the following defining properties:

- Large number of nodes with high spatial density
- Short range links with low data rate
- Low power, low complexity, low cost, small form factor
- Heterogeneous complexity constraints

9

These properties are typical for applications from body area networks [51] up to control of industrial production technology [52], networks-on-chip [53] or nano communications [54]. The first item is mainly a consequence of the recent advantages in very-large-scale integration (VLSI) and microelectromechanical systems (MEMS). Due to efficient mass production and miniaturization of sensors, we can expect that future sensor networks will be composed of thousands or millions of nodes with an increasing number of nodes per volume. Therefore, we consider the sensor network to have a large number of nodes with high spatial density. The second item concerns the typical communication pattern of the nodes. Generally, we consider the communication links of sensor networks to be of short distance, i.e. rather in the order of centimeters than kilometers. We address environments such as indoor communication with many nodes per room, instead of considering long range networks covering e.g. data links between cities. Furthermore, it is not the main objective of such sensor networks to provide very high data rates. The individual sensor nodes are considered to have a low duty cycle and transfer small blocks of data once in a while. This leads to a low data rate per node with a short communication distance due to the high node density. The third item addresses economic and environmental constraints, which are probably the most important to make wireless sensor networks competitive. The transceiver of the sensor nodes must be of low power consumption, low complexity and low cost. It is obvious that a profitable employment of a large sensor network requires low costs per node. This comes along with stringent complexity requirements and limitations on the size of the sensor nodes. The form factor is mainly affected by the size of the antenna. For mobile nodes, it is in particular important to have a low energy consumption. Without external power supply, the nodes must rely on batteries or power harvesting. Therefore, the sensor nodes need to be energy efficient. The last item addresses the asymmetric structure of sensor networks. Whereas the sensor nodes need to be of lowest complexity, the network may contain nodes with less stringent complexity requirements. This could be central units, which are stationary and connected to external power. Sometimes those nodes are also denoted as cluster heads, full function devices or readers. Usually we assume that the number of central units is small compared to the number of sensor nodes. Additionally, the central units may also be used for higher layer processing or as user interface. The complexity constraints of the physical layer signal processing may then be negligible compared to the complexity of the higher layers.

The relevance and the wide area of applications have trigged extensive research on wireless communication for sensor networks. Besides requirements on data rate and reliability, the main objectives are low power, low complexity and low costs. Many different systems and standards for the physical layer have been proposed. Most of the conventional systems are based on narrow-band communication. This is reasonable at first glance, because most hardware components are much easier to design for narrow-band systems such as antennas or amplifiers. From this point of view, the hardware is more complex the wider the bandwidth and it may sound paradox to propose UWB for wireless sensor networks. However, the advantages of UWB for sensor networks are more fundamental and the consequence of the following three reasons.

The first reason is the large frequency diversity of UWB. Conventional narrowband systems such as ZigBee or Wifi suffer severely from fading due to multipath propagation. Especially in environments with lots of metal, strong multipath components from scattering or reflections can be expected. We usually find harsh propagation environments in industrial production plants, machine rooms or around engines. The multipath is a problem for narrowband communication. Due to the use of almost periodical waveforms, the superposition of the inversely phased signals leads to cancelation, i.e. destructive interference. Thus, the signal quality fluctuates strongly over space or over time, e.g. when machine parts are moving. Much effort has been spend to combat fading in narrowband systems. The standard approaches are to collect diversity over space with multiple antennas or over time or frequency. Whereas the latter is difficult in narrowband systems, it is the inherent advantage of UWB. The large bandwidth of UWB with up to 7.5 GHz enables communication even in harsh propagation environments. With such a large bandwidth, there are always frequencies available with a good signal quality. In UWB systems, the transmit pulses are so short, that single multipath reflections can be resolved at the receiver. The signal components from multipath propagation do not overlap and therefore they cannot cancel each other. Even when the line-of-sight (LOS) path is shadowed, the signal energy from the multipath components can be collected at the receiver to decode the transmit data. The pulses are so short that the periodic parts are small. Hence, UWB systems are much less vulnerable to deep fades due to destructive interference compared to narrowband systems.

The second reason concerns the signal processing and is a consequence of the trade-off between spectrum efficiency and complexity. For narrowband systems, the scarcest and most valuable resource is bandwidth. Traditionally, the major design goal is data rate, i.e. to transfer the maximal number of bps/Hz under a transmit power constraint. The consequence of this optimization is a high transceiver complexity. To achieve a good performance and a high data rate, expensive computations and complex processing is necessary at the transmitter as well as receiver. Furthermore, the analog components need to be of high quality such as low-noise amplifiers with linear characteristic, mixers with high dynamic range, small in-phase and quadrature (IQ) imbalance or high resolution analog-to-digital converter (ADC). This makes conventional narrowband transceiver power hungry and costly, when they need to support an adequate performance. However, the design criteria for UWB communication are much different, because bandwidth efficiency becomes insignificant. With an available bandwidth of up to 7.5 GHz, it is not necessary to be bandwidth efficient. When considering sensor networks, the required data rate is much smaller than the available bandwidth. This opens much room for optimization of UWB transceivers for low complexity and low power operation. The large bandwidth enables sufficient performance even of receivers that are suboptimal in terms of detection performance. Whereas the design of narrowband systems usually starts from the derivation of optimal detectors, in UWB we can start from efficient hardware designs. With the relaxation on the spectral efficiency, UWB facilitates low complexity system designs.

The third reason is that UWB enables high definition localization and imaging as a secondary use of the communication system. Localization of sensor nodes is of particular importance for large networks. With millions of nodes, manual placement and identification of sensor nodes becomes impractical and for many cases the sensor data without location is useless. To include global navigation satellite system (GNSS) receivers in every sensor node is usually not an option, because their accuracy is low in indoor environments. For sensor networks, standard approaches from literature distinguish between fine-grained and coarse-grained localization. An overview on cooperative localization for sensor networks is given e.g. in [34, 36, 55]. Fine-grained localization is usually based on time of arrival (ToA) or angle of arrival (AoA) and coarse-grained localization on received signal strength (RSS) or other connectivity metrics. For fine-grained localization, there are two reasons why a large bandwidth is beneficial. First, the accuracy of timing estimation depends on the pulse duration. The larger the bandwidth, the shorter the pulse and the better the timing estimate in additive noise. A short pulse is steeper and thus the timing can be estimated more accurately. This leads to better performance of ToA or AoA based algorithms. The second reason is the resolvable multipath. With high bandwidth, the LOS path can be distinguished better from multipath components. This is important for harsh propagation environments, where narrowband localization would fail due to multipath. Also for coarse-grained localization, the high bandwidth is beneficial. Due to less small scale fading the RSS is more stable. Furthermore, innovative approaches such as location fingerprinting based on the shape of the channel impulse response can be used [39].

2.1 Non-coherent Ultra-Wideband Systems

The stringent requirements on complexity and power consumption prohibit the use of standard receivers for UWB sensor networks. The matched filter receiver or rake receiver are complex to implement for multipath channels with large bandwidth. The adaptation to the channel requires huge complexity in order to obtain the optimal decoding performance. Both digital and analog implementations are expensive and power hungry. On the one hand, a digital implementation of a channel matched filter requires high sampling rates and long filter lengths. With up to 7.5 GHz bandwidth, the receiver would require an ADC with at least 15 GHz sampling rate. This is not feasible in low complexity devices. On the other hand, an analog design is difficult to adapt to the channel. A rake receiver requires a large number of fingers to combine the signal contributions from multipath [56]. Both are not an option for low complexity nodes. Additionally, the channel estimation and dissemination of CSI requires high complexity, is expensive and consumes high power. The number of variables for channel estimation grows with increasing bandwidth.

The key to low complexity and low power consumption for UWB sensor networks are suboptimal receivers that can be implemented efficiently. The most prominent representatives for this approach are non-coherent UWB systems. For narrowband systems, non-coherent communication usually means that the phase of the channel is unknown and neglected at the receiver. For UWB, non-coherent communication refers commonly to a system with a receiver that does not require any knowledge of the channel. Although this comes with a loss in detection performance, it leads to hardware-efficient receiver designs. An elegant implementation of this concept is the energy detector and the autocorrelation receiver. The receivers can be build up with low complexity, particularly when binary modulation schemes are used. Fig. 2.1 shows the block diagrams of the receivers and the corresponding transmit signals.

The energy detection receiver is usually used in conjunction with binary and orthogonal modulation schemes such as BPPM or on-off-keying (OOK). With the latter one, the transmitter sends a pulse for transmit symbol $a = $ "1" and no pulse for $a = $ "0". The energy detection receiver computes the integral of the squared and bandpass filtered receive signal, i.e. it estimates the received signal energy over a certain time-window. This value is larger if a pulse is present and smaller without pulse. For OOK, the receiver decides for "1" for a large value and for "0" for a small value. However, it requires a threshold for the decision, which needs to be adapted depending on the received signal power and is difficult to implement. This problem is circumvented by the use of BPPM, see Fig 2.1a. Here, the symbol

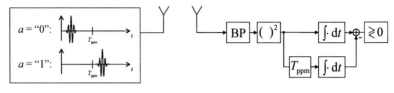

a) Binary Pulse Position Modulation   b) Energy Detection Receiver

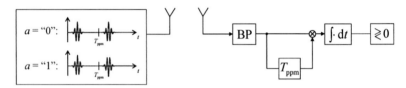

c) Transmitted Reference Modulation   d) Autocorrelation Receiver

Fig. 2.1: Conventional approaches for non-coherent UWB systems

is split into two time slots of length T_{ppm} each. A pulse is transmitted in the first time slot for $a =$ "0" and in the second for symbol $a =$ "1". With BPPM, the receiver estimates the energy that is received in both time slots and compares their values, as shown in Fig 2.1b. The upper branch corresponds to the first time slot and the lower the second time slot. Note that the box with T_{ppm} depicts a delay. The output of the integration over a fixed time interval is subtracted and can thus be compared to threshold zero. For positive values the receiver decides for $a =$ "0" and for negative values for $a =$ "1".

The other well-known receiver structure for non-coherent UWB is the autocorrelation receiver as shown in Fig. 2.1d. For this structure, the transmitter usually employs transmitter reference modulation, see Fig. 2.1c. Here, a pulse is always transmitted in the first timeslot, which serves as a reference for the second time slot. Whereas for $a =$ "0", the second time slot contains the same pulse, for $a =$ "1" the second pulse is sent with opposite polarity or phase. The receiver computes the correlation of the first and second time slot. This is done by the integration of the product of the signal of the first and second time slot. If the channel remains constant over both time slots, the correlation is maximal for $a =$ "0" and minimal for $a =$ "1". Thus, the sign of the output can be used for symbol decision. The performance

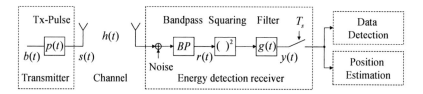

Fig. 2.2: Block diagram of transmitter, channel and generalized energy detection receiver

of the autocorrelation receiver is similar to the energy detection receiver [37] and it also does not require any channel knowledge. However, the implementation requires a perfect delay element of at least the channel excess delay length. Whereas this can easily be realized with a cable, it is hard to be integrated in VLSI. Therefore, we focus in this thesis on a hardware design which is based on the energy detection receiver. However, many of the proposed approaches can be extended to transmitted reference in a straightforward way.

2.2 Generalized Energy Detection Receiver

The generalized energy detection receiver is a particular elegant implementation of a non-coherent UWB receiver. It is depicted in Fig. 2.2, which shows the system model that is considered throughout this thesis. The system consists of the three basic parts: the UWB transmitter (TX), the wireless channel, and the generalized energy detection receiver (RX). The output of the receiver is used for data detection as well as position estimation.

The central element of the system is the wireless channel. We model the channel as a linear time invariant (LTI) system, which assumes that transmitter and receiver are stationary during a transmission. In this thesis, it is one of the key approaches to link the channel to the position of the transmitter and receiver. If the complete environment of the transmitter and receiver were known, the channel impulse response would be deterministic and could be predicted by Maxwell's equations. However, this would require the knowledge of the precise position and parameter of each particle in the environment. This is unrealistic, besides that it leads to complex equations that are difficult to solve. In practice, already the measurements of the position of the transmitter and receiver are affected by an error. It is one of the standard methods in engineering to tackle such a problem by considering a stochastic model instead of the deterministic model. Instead of assuming that everything is perfectly known, the environment and the channel are subject to uncertainty. We do not take the precise environment into account but rather consider the propagation effects to happen with a certain probability. This is incorporated in the system model by treating the channel impulse response $h(t)$ as a realization of a real-valued non-stationary random process

$$(H(t), t \in \mathbb{R}).$$

The distribution of $(H(t))$ depends on the environment and on the position of the transmitter and receiver. Specifically, we consider the distribution of the channel conditioned on the position of the transmitter $\mathbf{p}_{\text{tx}}$ and the position of the receiver $\mathbf{p}_{\text{rx}}$. The complete statistical description of the random process involves knowledge of the cumulative distribution function (CDF)

$$F_n(h_1, \ldots, h_n; t_1, \ldots, t_n | \mathbf{p}_{\text{tx}}, \mathbf{p}_{\text{rx}}) = \Pr\left[H(t_1) \leq h_1, \ldots, H(t_n) \leq h_n | \mathbf{p}_{\text{tx}}, \mathbf{p}_{\text{rx}}\right] \qquad (2.1)$$

for every $n \in \mathbb{N}$ and for every $t_1, \ldots, t_n \in \mathbb{R}$. We refer to (2.1) as *location-aware channel knowledge* or *regional channel knowledge*. The latter assumes that $\mathbf{p}_{\text{tx}}$ and $\mathbf{p}_{\text{rx}}$ are quantized to different regions.

Whereas the properties of the channel are given by nature, the transmitter and receiver are subject to the system design. For the data transmission, the transmitter uses BPPM. This is implemented by the following input signal:

$$b(t) = \sum_n c_n \delta(t - a_n T_{\text{ppm}} - n T_{\text{symb}}),$$

where $a_n \in \{0, 1\}$, $n \in \mathbb{Z}$ denotes the sequence of transmit symbols. The length of the time slot T_{ppm} is usually half the symbol duration T_{symb}. The polarity $c_n \in \{-1, 1\}$ of the pulses is chosen randomly to avoid discrete spectral lines in the transmit spectrum. With the pulse shape denoted by $p(t)$, the signal that is fed to the antenna can be written as

$$s(t) = p(t) * b(t).$$

This signal is sent over the wireless channel with impulse response $h(t)$. At the receiver, the signal is perturbed by a realization of additive white Gaussian noise (AWGN) with power spectral density $N_0/2$. The receiver input is limited by a bandpass filter with bandwidth B and center frequency f_c. For notational convenience, we generally include the bandpass filter in the impulse response of the channel and in the noise. The bandpass filtered noise is denoted by $n(t)$ and the band-limited channel impulse response by $\tilde{h}(t)$. Thus, we have

$$r(t) = \tilde{h}(t) * s(t) + n(t).$$

The band-limited channel impulse response $\tilde{h}(t)$ is as a realization of the random process $\left(\tilde{H}(t), t \in \mathbb{R} \right)$ with conditional distribution function

$$F(\tilde{h}(T), \ldots, \tilde{h}(NT) | \mathbf{p}_{\text{tx}}, \mathbf{p}_{\text{rx}}) = \Pr \left[\tilde{H}(T) \leq \tilde{h}(T), \ldots, \tilde{H}(NT) \leq \tilde{h}(NT) | \mathbf{p}_{\text{tx}}, \mathbf{p}_{\text{rx}} \right],$$

where $T \leq \frac{2}{f_c} + \frac{1}{B}$, $N \in \mathbb{N}$. In general, the exact description requires an infinite number of samples N. Here, however, we assume a finite number N, i.e. we approximate the band-limited channel impulse response to be time-limited. With the fixed sampling instances $T, \ldots, NT$, the conditional PDF of $\left(\tilde{H}(t), t \in \mathbb{R} \right)$ is then given by

$$p(\tilde{h}(T), \ldots, \tilde{h}(NT) | \mathbf{p}_{\text{tx}}, \mathbf{p}_{\text{rx}}) = \frac{\partial^N F(\tilde{h}(T), \ldots, \tilde{h}(NT) | \mathbf{p}_{\text{tx}}, \mathbf{p}_{\text{rx}})}{\partial \tilde{h}(T) \ldots \partial \tilde{h}(NT)}.$$

The signal processing of the generalized energy detection receiver consists of a squaring device and the post-detection filter with impulse response $g(t)$. The T_s-spaced samples at the output of the generalized energy detection receiver are given by

$$y(kT_s) = \int\limits_{-\infty}^{\infty} g(\tau) r^2(kT_s - \tau) \mathrm{d}\tau.$$

The conventional energy detector as shown in Fig. 2.1b is a special case of the generalized energy detection receiver. If the post-detection filter has a rectangular impulse response (sliding window integrator), the generalized energy detection receiver corresponds to a conventional energy detector:

$$y(kT_s) = \int\limits_{kT_s - T_{\text{int}}}^{kT_s} r^2(t) \mathrm{d}t,$$

where T_{int} denotes the length of the integration. The choice of the post-detection filter has large impact on the complexity as well as performance of the receiver. Instead of the rectangular filter, it may be preferable to use other filters:

- The time-reversed and squared receive pulse, i.e. $g(-t) = \left(p(t) * \tilde{h}(t) \right)^2$. This corresponds to the optimal filter under the approximation that $r^2(t)$ is Gaussian distributed.

- The time-reversed average power delay profile (APDP) as proposed in [57]. The generalized energy detection receiver coincides then with the maximum likelihood (ML) detector under the assumption that the channel impulse response is a white Gaussian random process with zero mean and covariance according to the APDP.

- A first order low-pass filter as shown in [58]. In terms of hardware complexity and power consumption, a simple first order low-pass filter outperforms the sliding window integrator.

The advantage of the generalized energy detection receiver is that the complexity and performance can be adjusted and smart trade-offs can be found. Depending on the requirements, the post-detection filter can be optimized for detection performance or complexity.

Chapter 3

Optimized Data Transmission

The following two chapters compose the first part of this thesis and focus on data transmission for UWB sensor networks. In this chapter, we study the location-aware UWB communication with generalized energy detection receivers. The presented results can be summarized as follows:

- We derive an explicit expression for the SINR for the generalized energy detection receiver as a function of the transmit pulse (Proposition 1) and the post-detection filter (Proposition 2).

- Optimization algorithms for the transmitter and receiver are deduced. As reference we first study a scenario in which full CSI is available for the optimization. Subsequently, we relax this condition and assume that only location (region) specific statistical CSI is available. We use a bit error rate (BER) criterion as cost function for the optimization of $g(t)$ and $p(t)$ and introduce some approximations that reduce the computational complexity.

- The performance of the proposed schemes is investigated on the basis of channel models and channel measurements.

Parts of this chapter have been published in [46].

3.1 Location-aware Adaptation and Precoding

This section introduces location-aware communication for UWB sensor networks. For many applications, joint localization and data communication is desirable, e.g. tracking items in a production hall, airport or hospital, combined with sensor data querying. If the location information is available, it can be used for performance enhancement of data transmission.

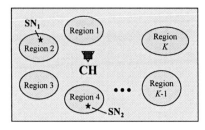

Fig. 3.1: Network scenario with a cluster head (CH), sensor nodes (SN), and regions $1, \ldots, K$.

So far, this approach has been investigated for higher layers such as routing [32] or security protocols [33]. We propose to use the location information on the physical layer to improve the performance of low complexity and low data rate UWB communication.

As an example, let us consider the following scenario: As shown in Fig. 3.1, the coverage area of a network might be divided into K subregions. We consider a heterogeneous network with low complexity UWB sensor nodes (SNs). These reduced function devices communicate with stationary cluster heads (CHs), which are full-function devices with less stringent complexity constraints. The downlink (CH to SNs) is the bottleneck in this scenario since the SNs are of low complexity and achieve a lower receiver sensitivity than the CH. Therefore, we consider the optimization of the downlink data transmission in the following discussion.

Conventional approaches attempt to adapt the transceiver directly to the channel state. They require the knowledge of the channel state over the full transmission bandwidth. Due to the huge bandwidth of UWB, the estimation and dissemination of CSI requires high complexity and is very expensive and power hungry. To circumvent this problem, we propose to adapt the transceiver to the region of the SN. This is done by modeling the channel impulse response as a random process with region dependent parameters, which can be estimated in an off-line training phase. The data transmission is then optimized based on the statistical channel knowledge for each region. This is the key to location-aware communication. The CHs are assumed to be able to determine the region, in which a SN is located. Based on the statistical channel knowledge for each region, we propose two different approaches to improve the data transmission:

- Transmitter optimization at CH: The CH uses a region specific transmit signal depending on where the SN is located. This approach uses a fixed post-detection filter receiver at the SN and applies precoding at the CH according to the channel statistics of the region.

- Receiver optimization at SN: This approach uses adaptable post-detection filters at the SN. Depending on its position or region, the SN can adjust the post-detection filter accordingly without requiring channel estimation.

Both approaches improve the detection performance and can be used to successfully suppress the influence of narrowband interference.

Related Work: To the best of our knowledge, precoding for UWB generalized energy detection receivers has so far not been considered in literature. Transmitter optimization and pulse shaping for coherent receivers has been extensively studied in literature, e.g. [59–61]. Common objectives are spectral mask, bit error rate [59] or multiuser performance [60] as well as narrowband interference suppression for linear receivers [61]. For low-complexity receivers, time-reversal prefiltering [62–64] and channel-phase-precoded (CPP) UWB [65,66] have been proposed. In [67], pre-equalization with spectral mask constraints is presented. Even though these precoding methods are related to the one we propose in this work, they neither suppress narrowband interference nor are they suitable for generalized energy detection receivers. With regard to receiver optimization, our work is related to [57, 68–70]. In [57], the ML decision rule based on the average PDP of the channel impulse response is derived and [69] presents ML detectors with partial CSI and intersymbol interference. Ref. [70] evaluates the performance of these receivers in a body area network environment. In [68], energy detection with multiple energy measurements is studied. However, compared to the approaches reported in literature, we take also narrowband interference into account and explicitly impose a stringent constraint on the receiver complexity.

3.2 System Model

In the following, we look at the transmission of a single symbol $a_0 \in \{0,1\}$. The considered transmission scheme can be modeled in discrete time[1] and vector notation as depicted in Fig. 3.2. The upper branch corresponds to the first time slot and the lower branch to the second time slot.

The transmit pulse is denoted by $\mathbf{p} \in \mathbb{R}^N$ with

$$\mathbf{p} = [p(T), \ldots, p(NT)]^T,$$

---

[1]Note that this sampling is only necessary for the theoretical analysis. The receiver implementation can be in analog without any high rate sampling.

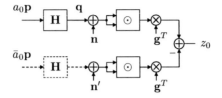

Fig. 3.2: Discrete time signal model

where T denotes the sampling period. The receive pulse $\mathbf{q} \in \mathbb{R}^N$ can be written as

$$\mathbf{q} = \mathbf{H}\mathbf{p},$$

where the channel matrix $\mathbf{H}$ includes the bandpass filter and has a Toeplitz structure with dimensions $N \times N$. It collects the shifted and zero-padded versions of the bandpass filtered channel impulse response $\tilde{\mathbf{h}} = \left[\tilde{h}(T), \ldots, \tilde{h}(N_hT), 0, \ldots, 0\right]^T$ on its columns, where N_h times the sampling period denotes the channel excess delay:

$$\mathbf{H} = \begin{bmatrix} \tilde{h}(T) & 0 & \cdots & \\ \tilde{h}(2T) & \tilde{h}(T) & 0 & \ddots \\ \vdots & \tilde{h}(2T) & \tilde{h}(T) & \\ \tilde{h}(N_hT) & \vdots & \tilde{h}(2T) & \ddots \\ 0 & \tilde{h}(N_hT) & \vdots & \ddots \\ \vdots & 0 & \tilde{h}(N_hT) & \\ & \vdots & \ddots & \ddots \end{bmatrix}$$

Thermal noise plus narrowband interference are modeled as zero mean Gaussian random vectors $\mathbf{n}$ (first time slot) and $\mathbf{n}'$ (second time slot). Their covariance matrices including the bandpass filter are given by

$$\boldsymbol{\Sigma}_{nn} = \mathsf{E}[\mathbf{n}\mathbf{n}^T] = \mathsf{E}[\mathbf{n}'\mathbf{n}'^T] = \boldsymbol{\Sigma}_{\tilde{w}} + \boldsymbol{\Sigma}_{\tilde{i}}$$

and

$$\boldsymbol{\Sigma}_{n'n} = \mathsf{E}[\mathbf{n}\mathbf{n}'^T] = \mathsf{E}[\mathbf{n}'\mathbf{n}^T]^T.$$

$\boldsymbol{\Sigma}_{\tilde{w}}$ and $\boldsymbol{\Sigma}_{\tilde{i}}$ denote the covariance matrix of the band-limited noise and interference, respectively. For notational convenience, the impulse response of the post-detection filter is stacked into a vector in reverse order, i.e.

$$\mathbf{g} = [g(NT), \ldots, g(T)]^T.$$

The matrix $\mathbf{G} = \text{diag}(\mathbf{g})$ carries these values on the main diagonal.

This system model is equivalent to the continuous time transceiver structure depicted in Fig. 2.2 under the following assumptions:

- The sampling period T must fulfill at least $\frac{1}{T} \geq 2B + 4f_c$ to account for the squaring operation.

- Edge effects due to the finite length of the signals are negligible.

- Intersymbol interference can be neglected, i.e.,

$$\int_{-\infty}^{\infty} g(\tau)q^2(t-\tau)d\tau = 0 \quad \text{for} \quad t \notin [0, T_{\text{ppm}}],$$

where $q(t)$ denotes the received pulse. This condition is fulfilled if the channel excess delay plus the time duration of the transmit pulse and impulse response of the post-detection filter is smaller than the PPM half slot T_{ppm}. This is a reasonable assumption in our context, because in sensor networks the per node peak data rate is moderate.

- Narrowband interference is Gaussian distributed.

- Furthermore, we require in the following, that the transmit bits are equally probable and only one pulse is transmitted per symbol with perfect synchronization at the receiver.

3.3 Signal-to-Interference-and-Noise Ratio

The SINR is defined as

$$\text{SINR} = \frac{\text{E}[z_0|a_0=0]^2}{\text{Var}[z_0|a_0=0]} = \frac{\text{E}[z_0|a_0=1]^2}{\text{Var}[z_0|a_0=1]}, \tag{3.1}$$

where z_0 denotes the receiver output and expectation is with respect to noise and interference. Conditioned on the transmit symbol, the SINR corresponds to the squared mean of the decision variable over its variance. Since the mean and the variance are the same for both transmit symbols and they occur with equal probability, the SINR is independent of the transmit symbol. Therefore, it is sufficient to consider the case that the pulse is in the first time-slot, i.e. $a_0 = 0$. The decision variable z_0 is then given by

$$z_0|_{a_0=0} = \mathbf{g}^T \left[(\mathbf{q} + \mathbf{n}) \odot (\mathbf{q} + \mathbf{n}) \right] - \mathbf{g}^T \left[\mathbf{n}' \odot \mathbf{n}' \right]. \tag{3.2}$$

Expanding the squares leads to

$$z_0|_{a_0=0} = \underbrace{\mathbf{g}^T (\mathbf{q} \odot \mathbf{q})}_{=:\alpha} + \underbrace{2\mathbf{g}^T (\mathbf{q} \odot \mathbf{n})}_{=:\beta} + \underbrace{\mathbf{g}^T (\mathbf{n} \odot \mathbf{n}) - \mathbf{g}^T (\mathbf{n}' \odot \mathbf{n}')}_{=:\gamma}.$$

The term α collects the desired signal contribution, i.e. the receive pulse $\mathbf{q}$, which is squared and convolved with the post-detection filter. All perturbing contributions are combined in the other terms: in β the product of the signal and the noise plus interference, and in γ the squared noise plus interference contribution of the first and second timeslot, respectively.

The SINR can then be written in terms of the squared signal, the mixed term and the squared noise term as follows:

$$\text{SINR} = \frac{\mu_\alpha^2}{\sigma_\beta^2 + \sigma_\gamma^2}$$

The numerator μ_α^2 represents the desired squared signal contribution α and the denominator is given by the variance σ_β^2 of the mixed signal and interference plus noise term and the variance σ_γ^2 of the squared noise and interference γ.

This signal structure at the energy detector output leads to the particular characteristic of this receiver structure. On the one hand, the squared terms are not Gaussian distributed anymore and can be shaped by the choice of the post-detection filter. On the other hand, the mixed term β depends on the transmit pulse as well. This enables distinct features of transmitter optimization, e.g. specific transmit pulse shapes for interference suppression.

Proposition 1. *The SINR in terms of the transmit pulse* $\mathbf{p}$ *is given by*

$$\text{SINR} = \frac{\left(\mathbf{p}^T \mathbf{A} \mathbf{p} \right)^2}{\mathbf{p}^T \mathbf{B}_\beta \mathbf{p} + \sigma_\gamma^2}, \tag{3.3}$$

where

$$\mathbf{A} = \mathbf{H}^T \mathbf{G} \mathbf{H},$$
$$\mathbf{B}_\beta = 4\mathbf{H}^T \mathbf{G} \mathbf{\Sigma}_{nn} \mathbf{G}^T \mathbf{H}$$
$$\sigma_\gamma^2 = 4\mathbf{g}^T \left(\mathbf{\Sigma}_{nn} \odot \mathbf{\Sigma}_{nn} - \mathbf{\Sigma}_{n'n} \odot \mathbf{\Sigma}_{n'n} \right) \mathbf{g}.$$

Proof. First, we compute the numerator of (3.1). The expectation of (3.2) yields

$$\mu_\alpha = \mathsf{E}\,[z_0|a_0 = 0] = \mathsf{E}\,[\alpha + \beta + \gamma] = \mathsf{E}[\alpha] = \alpha. \qquad (3.4)$$

The mixed signal and interference plus noise component β is zero mean and the means of the squared interference plus noise components γ cancel out, since the random processes are assumed to be stationary. It follows that

$$\alpha = \mathbf{g}^T \left[(\mathbf{Hp}) \odot (\mathbf{Hp}) \right]$$
$$= \sum_{k=1}^{N} g[N-k](\mathbf{p}^T \mathbf{h}_k \mathbf{h}_k^T \mathbf{p}), \qquad (3.5)$$

where $\mathbf{h}_k^T$ denotes the kth row of $\mathbf{H}$. Expression (3.5) can be further simplified to

$$\mathbf{p}^T \left(\sum_{k=1}^{N} g[N-k] \mathbf{h}_k \mathbf{h}_k^T \right) \mathbf{p} = \mathbf{p}^T \mathbf{H}^T \mathbf{G} \mathbf{H} \mathbf{p}. \qquad (3.6)$$

For the denominator, the variance of (3.2) yields

$$\mathsf{Var}[z_0|a_0 = 0] = \mathsf{E}\left[(z_0 - \mu_\alpha)^2 \right]$$
$$= \underbrace{\mathsf{E}\left[\beta^2\right]}_{=:\sigma_\beta^2} + \underbrace{\mathsf{E}\left[\gamma^2\right]}_{=:\sigma_\gamma^2} + \underbrace{\mathsf{E}\,[2\beta\gamma]}_{=0}.$$

The second moment of β computes to

$$\sigma_\beta^2 = \mathsf{E}\left[\left(2\mathbf{g}^T \left[(\mathbf{Hp}) \odot \mathbf{n} \right] \right)^2 \right]$$
$$= 4\mathsf{E}\left[((\mathbf{Gn})^T \mathbf{Hp})^2 \right]$$
$$= \mathbf{p}^T 4\mathbf{H}^T \mathbf{G} \underbrace{\mathsf{E}\left[\mathbf{nn}^T \right]}_{=\mathbf{\Sigma}_{nn}} \mathbf{G}^T \mathbf{Hp}.$$

25

With $E[\gamma] = 0$ follows for the squared interference plus noise term

$$\sigma_\gamma^2 = E\left[\left(\mathbf{g}^T(\mathbf{n}\odot\mathbf{n})\right)^2 - 2\mathbf{g}^T(\mathbf{n}\odot\mathbf{n})\mathbf{g}^T(\mathbf{n}'\odot\mathbf{n}') + \left(\mathbf{g}^T(\mathbf{n}'\odot\mathbf{n}')\right)^2\right]$$
$$= 2\mathbf{g}^T E\left[(\mathbf{n}\odot\mathbf{n})(\mathbf{n}\odot\mathbf{n})^T\right]\mathbf{g} - 2\mathbf{g}^T E\left[(\mathbf{n}\odot\mathbf{n})(\mathbf{n}'\odot\mathbf{n}')^T\right]\mathbf{g}$$
$$= 4\mathbf{g}^T\left(\boldsymbol{\Sigma}_{nn}\odot\boldsymbol{\Sigma}_{nn} - \boldsymbol{\Sigma}_{n'n}\odot\boldsymbol{\Sigma}_{n'n}\right)\mathbf{g}. \tag{3.7}$$

The last line follows from the property of the zero mean multivariate normal distribution [71]

$$E\left[n_i^2 n_j^2\right] = 2E\left[n_i n_j\right]^2 + E\left[n_i^2\right]E\left[n_j^2\right],$$

where n_i and n_j denotes the i-th and j-th element of the vector $\mathbf{n}$, respectively. Collecting terms and substitution of $\mathbf{A}$ and $\mathbf{B}_\beta$ yields finally (3.3). $\qquad\square$

Proposition 2. *The SINR in terms of the post-detection filter* $\mathbf{g}$ *is given by*

$$\text{SINR} = \frac{\mathbf{g}^T\mathbf{K}\mathbf{g}}{\mathbf{g}^T\mathbf{C}\mathbf{g}} \tag{3.8}$$

where

$$\mathbf{K} = \left(\mathbf{H}\mathbf{p}\mathbf{p}^T\mathbf{H}^T\right)\odot\left(\mathbf{H}\mathbf{p}\mathbf{p}^T\mathbf{H}^T\right)$$
$$\mathbf{C} = 4\left[\left(\mathbf{H}\mathbf{p}\mathbf{p}^T\mathbf{H}^T\right)\odot\boldsymbol{\Sigma}_{nn} + \boldsymbol{\Sigma}_{nn}\odot\boldsymbol{\Sigma}_{nn} - \boldsymbol{\Sigma}_{n'n}\odot\boldsymbol{\Sigma}_{n'n}\right].$$

Proof. With (3.4) we obtain for the numerator of (3.1)

$$\mu_\alpha^2 = \left(\mathbf{g}^T(\mathbf{q}\odot\mathbf{q})\right)^2$$
$$= \mathbf{g}^T(\mathbf{q}\odot\mathbf{q})(\mathbf{q}\odot\mathbf{q})^T\mathbf{g}$$
$$= \mathbf{g}^T\left[\left(\mathbf{q}\mathbf{q}^T\right)\odot\left(\mathbf{q}\mathbf{q}^T\right)\right]\mathbf{g}$$
$$= \mathbf{g}^T\left[\left(\mathbf{H}\mathbf{p}\mathbf{p}^T\mathbf{H}^T\right)\odot\left(\mathbf{H}\mathbf{p}\mathbf{p}^T\mathbf{H}^T\right)\right]\mathbf{g}.$$

For the denominator of (3.1), we note that σ_β^2 can be written in terms of $\mathbf{g}$ as

$$\sigma_\beta^2 = 4E\left[\left(\mathbf{g}^T(\mathbf{q}\odot\mathbf{n})\right)^2\right]$$
$$= 4\mathbf{g}^T E\left[(\mathbf{q}\odot\mathbf{n})(\mathbf{q}\odot\mathbf{n})^T\right]\mathbf{g}$$
$$= 4\mathbf{g}^T\left[\left(\mathbf{q}\mathbf{q}^T\right)\odot E\left[\mathbf{n}\mathbf{n}^T\right]\right]\mathbf{g}$$
$$= \mathbf{g}^T\left[4\left(\mathbf{H}\mathbf{p}\mathbf{p}^T\mathbf{H}^T\right)\odot\boldsymbol{\Sigma}_{nn}\right]\mathbf{g}.$$

With σ_γ^2 as given in (3.7) follows directly (3.8).   □

3.4 Optimization of Transmitter or Receiver

This section presents optimization algorithms based on the SINR expressions that have been derived in Section 3.3. First, we present the optimization of the transmit pulse **p** and post-detection filter **g** for a given channel realization. This provides performance bounds for the generalized energy detection receiver. Such a transmission scheme requires the full knowledge of the channel impulse response at the transmitter and/or receiver, which might be expensive to obtain and might require a lot of overhead. Hence, in the second part of this section, we present optimization schemes based on statistical channel knowledge, i.e. the covariance matrix of the channel impulse response to incorporate the location knowledge.

3.4.1 Full channel knowledge

Transmitter Optimization: Given a certain post-detection filter, the optimization problem for the transmit pulse can be formulated as follows:

$$\mathbf{p}^* = \underset{\mathbf{p} \in \mathbb{R}^N : \|\mathbf{p}\| = 1}{\arg\max} \ \text{SINR} \tag{3.9}$$

The maximization is subject to a norm constraint on **p**, i.e. the energy per bit is limited. For environments where the FCC regulation [11] is applied, the given constraint is an approximation on the peak level of emissions. According to the FCC, the peak power is restricted to 0 dBm after filtering with 50 MHz centered at the frequency with the highest emission. Since the scope of this chapter is on low data rate systems, the average power constraint is without effect [72]. It restricts the power to -41.3 dBm/MHz contained on a 1 MHz band when averaged over 1 ms. For sparse transmit signals with a low duty cycle, the average power is small, due to the long averaging over 1 ms. However, to extend the optimization to high data rate systems, the average power constraint could also be included, e.g. as presented in [67]. Optionally, the transmit pulse must be scaled to be compliant with the regulation.

With the substitution

$$\mathbf{p} \mapsto \frac{\tilde{\mathbf{p}}}{\|\tilde{\mathbf{p}}\|},$$

the norm power constraint is fulfilled for all $\tilde{\mathbf{p}} \in \mathbb{R}^N$. With (3.3) from Proposition 1, we can write (3.9) then as an unconstrained optimization problem

$$\mathbf{p}^* = \underset{\tilde{\mathbf{p}} \in \mathbb{R}^N}{\arg\max} \frac{\left(\tilde{\mathbf{p}}^T \mathbf{A} \tilde{\mathbf{p}}\right)^2}{\tilde{\mathbf{p}}^T \mathbf{B} \tilde{\mathbf{p}} \cdot \tilde{\mathbf{p}}^T \tilde{\mathbf{p}}}, \tag{3.10}$$

where

$$\mathbf{B} = 4\mathbf{H}^T \mathbf{G} \mathbf{\Sigma}_{nn} \mathbf{G}^T \mathbf{H} + 4(\mathbf{g}^T (\mathbf{\Sigma}_{nn} \odot \mathbf{\Sigma}_{nn} - \mathbf{\Sigma}_{n'n} \odot \mathbf{\Sigma}_{n'n}) \mathbf{g})\mathbf{I}.$$

Standard numerical algorithms can be used to solve this optimization problem. In Section 3.4.2 we provide an approach based on the Newton's method.

Low SINR Approximation (LSA): Depending on the operating regime of the receiver, the influence of the mixed signal and interference plus noise component and squared interference plus noise component is different. If the received signal power is high, the mixed term dominates in the denominator of the SINR expression. However, for a rather low SINR, which is the operating regime of typical UWB-IR communication, the squared interference plus noise is much larger than the mixed term, i.e. $\sigma_\beta^2 \ll \sigma_\gamma^2$. When the mixed term in the denominator is omitted, this corresponds to a different SINR definition:

$$\text{SINR}_{\text{LSA}} = \frac{\text{Var}\left[z_0 | \mathbf{n} = 0, \mathbf{n}' = 0\right]}{\text{Var}\left[z_0 | \mathbf{q} = 0\right]}$$

Note that this is in line with the signal-to-noise-ratio definition for coherent receivers. The signal component in the numerator is given by the variance of the decision variable without noise and the denominator accordingly by the variance without signal. Substituting these values, we obtain

$$\text{SINR}_{\text{LSA}} = \frac{\alpha^2}{\sigma_\gamma^2} = \frac{\left(\mathbf{p}^T \mathbf{H}^T \mathbf{G} \mathbf{H} \mathbf{p}\right)^2}{4\mathbf{g}^T (\mathbf{\Sigma}_{nn} \odot \mathbf{\Sigma}_{nn} - \mathbf{\Sigma}_{n'n} \odot \mathbf{\Sigma}_{n'n}) \mathbf{g}}.$$

The maximization of SINR_{LSA} leads to an interesting approximation of the transmitter optimization problem:

$$
\begin{aligned}
\mathbf{p}^*_{\text{LSA}} &= \underset{\mathbf{p} \in \mathbb{R}^N : \|\mathbf{p}\| = 1}{\arg\max} \ \text{SINR}_{\text{LSA}} \\
&= \underset{\mathbf{p} \in \mathbb{R}^N : \|\mathbf{p}\| = 1}{\arg\max} \ \left(\mathbf{p}^T \underbrace{\mathbf{H}^T \mathbf{G} \mathbf{H}}_{=\mathbf{A}} \mathbf{p}\right)^2 = \underset{\tilde{\mathbf{p}} \in \mathbb{R}^N}{\arg\max} \ \left|\frac{\tilde{\mathbf{p}}^T \mathbf{A} \tilde{\mathbf{p}}}{\tilde{\mathbf{p}}^T \tilde{\mathbf{p}}}\right|.
\end{aligned} \tag{3.11}
$$

The solution to (3.11) is given by the principal eigenvector of $\mathbf{A}$, i.e.

$$\mathbf{p}_{LSA}^* = \mathbf{v}_{max}\{\mathbf{A}\}.$$

Receiver Optimization: Considering now the post-detection filter, the optimization problem for a fixed transmit pulse can be written as:

$$\mathbf{g}^* = \underset{\mathbf{g} \in \mathbb{R}^N}{\arg\max} \, \text{SINR} = \underset{\mathbf{g} \in \mathbb{R}^N}{\arg\max} \, \frac{\mathbf{g}^T \mathbf{K} \mathbf{g}}{\mathbf{g}^T \mathbf{C} \mathbf{g}} \tag{3.12}$$

With Proposition 2, the objective function results in a fraction of two quadratic forms. The solution to this optimization problem is directly given by

$$\mathbf{g}^* = \mathbf{v}_{max}\{\mathbf{K}, \frac{1}{2}(\mathbf{C} + \mathbf{C}^T)\}.$$

Joint Optimization of Transmitter and Receiver: So far, we considered optimization of transmitter or receiver for the case that the other part of the system is fixed. However, for the situation that transmitter as well as receiver are adaptable, it would be desirable to optimize transmitter and receiver jointly. Even though such a system configuration may not be realistic in a system with stringent complexity requirements, the joint optimization of $\mathbf{p}$ and $\mathbf{g}$ is also worthwhile to discover the performance bounds of the considered system. The joint optimization of transmitter and receiver is formulated by

$$\begin{bmatrix} \tilde{\mathbf{p}}^* \\ \mathbf{g}^* \end{bmatrix} = \underset{[\tilde{\mathbf{p}}^T, \mathbf{g}^T]^T \in \mathbb{R}^{2N}}{\arg\max} \, \text{SINR}.$$

The solution to the optimization problem is discussed in the next section.

3.4.2 Solution to Optimization Problems

The receiver optimization is directly given by a generalized eigenvector problem and standard algorithms can be used to find the solution. However, the objective function for transmitter and joint optimization is more complicated. Further analysis is of great use to solve the optimization problems.

Transmitter Optimization: The passband system model requires strong oversampling of the transmit pulse **p**. This is in particular due to the squaring of the signal. However, for the transmitter optimization the number of relevant dimensions is limited by the bandpass filter. Only the signal components that can pass the bandpass filter contribute to the symbol decision. For all other signal components, the numerator as well as denominator of the SINR becomes small. For the solution to the transmitter optimization we propose to reduce the dimension to the ones that are relevant. This has the advantage that i) the optimization problem can be solved faster and ii) the denominator of the objective function cannot become zero. To reduce to the relevant dimensions, the optimization problem can be written as

$$\mathbf{p}^* = \underset{\tilde{\mathbf{p}} \in \mathbb{R}^N}{\arg\max} \frac{\left(\tilde{\mathbf{p}}^T \mathbf{A} \tilde{\mathbf{p}}\right)^2}{\tilde{\mathbf{p}}^T \mathbf{B} \tilde{\mathbf{p}} \cdot \tilde{\mathbf{p}}^T \tilde{\mathbf{p}}} = \mathbf{V} \underset{\tilde{\mathbf{x}} \in \mathbb{R}^N}{\arg\max} \frac{\left(\tilde{\mathbf{x}}^T \mathbf{D} \tilde{\mathbf{x}}\right)^2}{\tilde{\mathbf{x}}^T \mathbf{V}^T \mathbf{B} \mathbf{V} \tilde{\mathbf{x}} \cdot \tilde{\mathbf{x}}^T \tilde{\mathbf{x}}},$$

where the diagonal of **D** and the columns of **V** are the eigenvalues and corresponding eigenvectors of **A**, respectively, i.e. $\mathbf{AV} = \mathbf{VD}$. The relevant dimensions are the eigenvectors with large eigenvalue. The eigenvectors with eigenvalue close to zero can be omitted. Their contribution can only increase the denominator, which is suboptimal under the given constraint. Let the diagonal of $\mathbf{D}_1$ be the N_{BP} non-zero eigenvalues and let the columns of $\mathbf{V}_1$ be the corresponding eigenvectors. The optimization problem then yields

$$\mathbf{p}^* = \mathbf{V}_1 \cdot \underset{\mathbf{x} \in \mathbb{R}^{N_{\mathrm{BP}}}}{\arg\max} \frac{\left(\mathbf{x}^T \mathbf{D}_1 \mathbf{x}\right)^2}{\mathbf{x}^T \mathbf{V}_1^T \mathbf{B} \mathbf{V}_1 \mathbf{x} \cdot \mathbf{x}^T \mathbf{x}}.$$

As an approximation, the eigenvectors with small eigenvalues (but not equal to zero) may also be neglected. The eigenvectors can be selected based on a threshold or based on their count. The number of relevant eigenvectors is approximately given by $N_{\mathrm{BP}} \approx 2BT_{\mathrm{ppm}}$. Assuming that the eigenvalues are sorted on the diagonal of **D** in descending order, we obtain $\mathbf{D}_1 = [\mathbf{D}]_{1:N_{\mathrm{BP}},1:N_{\mathrm{BP}}}$ and $\mathbf{V}_1 = [\mathbf{V}]_{1:N,1:N_{\mathrm{BP}}}$. Note that with $\mathbf{x} = [1,0,\ldots,0]^T$ we obtain the principle eigenvector, which is the solution for the low SINR approximation:

$$\mathbf{V}_1 \cdot [1,0,\ldots,0]^T = [\mathbf{V}_1]_{1:N,1} = \mathbf{v}_{\max}\{\mathbf{A}\} = \mathbf{p}_{\mathrm{LSA}}^*.$$

To solve the optimization problem in $\mathbf{x}$, we propose to apply Newton's Method. The update rule is given by

$$\mathbf{x}_{k+1} = \mathbf{x}_k - \mu \left[H_{\mathbf{x}_k} \text{SINR} \right]^{-1} \nabla_{\mathbf{x}_k} \text{SINR}, \tag{3.13}$$

where $\nabla_{\mathbf{x}_k} \text{SINR}$ denotes the gradient vector of the SINR with respect to $\mathbf{x}_k$, $H_{\mathbf{x}_k} \text{SINR}$ the Hessian matrix and μ the step size parameter. The gradient computes to

$$\nabla_{\mathbf{x}_k} \text{SINR} = \nabla_{\mathbf{x}_k} \frac{s^2}{n \cdot c} = \frac{2s}{n \cdot c} \nabla_{\mathbf{x}_k} s - \frac{s^2}{(n \cdot c)^2} (c \nabla_{\mathbf{x}_k} n + n \nabla_{\mathbf{x}_k} c) \tag{3.14}$$

where

$$s := \mathbf{x}_k^T \mathbf{D}_1 \mathbf{x}_k$$
$$n := \mathbf{x}_k^T \mathbf{V}_1^T \mathbf{B} \mathbf{V}_1 \mathbf{x}_k$$
$$c := \mathbf{x}_k^T \mathbf{x}_k.$$

and

$$\nabla_{\mathbf{x}_k} s = 2 \mathbf{D}_1 \mathbf{x}_k$$
$$\nabla_{\mathbf{x}_k} n = 2 \mathbf{V}_1^T \mathbf{B} \mathbf{V}_1 \mathbf{x}_k$$
$$\nabla_{\mathbf{x}_k} c = 2 \mathbf{x}_k.$$

The Hessian matrix with respect to $\mathbf{x}_k$ is given by

$$\begin{aligned} H_{\mathbf{x}_k} \text{SINR} =& \nabla_{\mathbf{x}_k} (\nabla_{\mathbf{x}_k} \text{SINR})^T \\ =& \mathbf{v}_1 (\nabla_{\mathbf{x}_k} s)^T - \mathbf{v}_2 (\nabla_{\mathbf{x}_k} n)^T - \mathbf{v}_3 (\nabla_{\mathbf{x}_k} c)^T \\ &+ \frac{2s}{n \cdot c} 2 \mathbf{D}_1 - \frac{s^2}{n^2 \cdot c} 2 \mathbf{V}_1^T \mathbf{B} \mathbf{V}_1 - \frac{s^2}{n \cdot c^2} 2 \mathbf{I} \end{aligned} \tag{3.15}$$

where

$$\mathbf{v}_1 := \frac{2}{nc} \nabla_{\mathbf{x}_k} s - \frac{2s}{(nc)^2} (c \nabla_{\mathbf{x}_k} n + n \nabla_{\mathbf{x}_k} c)$$
$$\mathbf{v}_2 := \frac{2s}{n^2 c} \nabla_{\mathbf{x}_k} s - \frac{s^2}{(n^2 c)^2} \left(c \cdot 2n \nabla_{\mathbf{x}_k} n + n^2 \nabla_{\mathbf{x}_k} c \right)$$
$$\mathbf{v}_3 := \frac{2s}{c^2 n} \nabla_{\mathbf{x}_k} s - \frac{s^2}{(c^2 n)^2} \left(c^2 \nabla_{\mathbf{x}_k} n + n \cdot 2c \nabla_{\mathbf{x}_k} c \right).$$

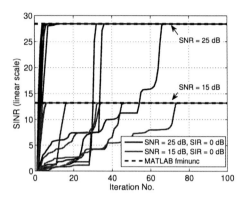

Fig. 3.3: Transmitter optimization: Convergence of Newton's Method

Due to the substitution of the power constraint, the SINR is independent of the norm of $\mathbf{x}_k$. Therefore, the Hessian matrix $H_{\mathbf{x}_k}$ SINR is rank deficient and cannot be inverted as required for the Newton's method. To circumvent this problem, we propose to fix the first element of $\mathbf{x}_k$ and to perform the optimization only with the remaining elements of $\mathbf{x}_k$. Then, we observed that the Hessian matrix has full rank and the Newton's method shows fast convergence.

As an example, Fig. 3.3 shows the trajectories of the objective function for randomly chosen starting vectors. The iterations are according to (3.13) with the step size parameter μ obtained by line search. For this example, we chose the following parameters, which are defined in detail in Section 3.5: $T_{\text{ppm}} = 50\,\text{ns}$, Bandwidth $B = 3\,\text{GHz}$, $N_{\text{BP}} = 300$, center frequency $f_c = 4.5\,\text{GHz}$, narrowband interference with bandwidth $\text{BW}_i = 10\,\text{MHz}$ at $f_i = 4.5\,\text{GHz}$, post-detection filter: first-order low-pass with $f_{\text{cutoff}} = 25\,\text{MHz}$.

Unfortunately, we cannot formally prove convergence of the optimization problem and to the best of our knowledge, this has not been considered in literature. However, in numerous simulations, it was observed that always the same result is approached. Even with a large number of realizations, we could not find any subsequence that converges to another value. Hence, we conjecture that the optimization is generally independent of the starting vector.

Joint Optimization of Transmitter and Receiver: So far, joint optimization of transmitter and receiver can easily be obtained by an alternating optimization. That is, the transmitter is optimized for a fixed receiver, the receiver for this transmitter and then the transmitter again, etc. However, this method converges slowly, see Fig 3.4. To improve the convergence,

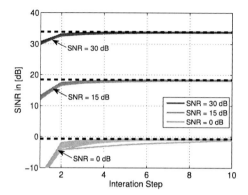

Fig. 3.4: Joint Optimization: Convergence of alternating optimization compared to final result from Newton's method (dashed lines)

we propose to use Newton's method to simultaneously optimize transmitter and receiver. The final result of this optimization is plotted with dashed lines in Fig 3.4. The update equation for the iterative algorithm is given by

$$\begin{bmatrix} \mathbf{x}_{k+1} \\ \mathbf{g}_{k+1} \end{bmatrix} = \begin{bmatrix} \mathbf{x}_k \\ \mathbf{g}_k \end{bmatrix} - \mu \left[H_{\mathbf{x}_k, \mathbf{g}_k} \text{SINR} \right]^{-1} \begin{bmatrix} \nabla_{\mathbf{x}_k} \text{SINR} \\ \nabla_{\mathbf{g}_k} \text{SINR} \end{bmatrix}.$$

The joint gradient is composed of the gradient with respect to $\mathbf{x}_k$ as given in (3.14) and the gradient with respect to $\mathbf{g}_k$, given by

$$\nabla_{\mathbf{g}_k} \text{SINR} = \frac{1}{\mathbf{g}_k^T \mathbf{C} \mathbf{g}_k} 2\mathbf{K}\mathbf{g}_k - \frac{\mathbf{g}_k^T \mathbf{K} \mathbf{g}_k}{\left(\mathbf{g}_k^T \mathbf{C} \mathbf{g}_k \right)^2} (\mathbf{C} + \mathbf{C}^T) \mathbf{g}_k.$$

The joint Hessian matrix computes to

$$H_{\mathbf{x}_k, \mathbf{g}_k} \text{SINR} = \nabla_{\mathbf{x}_k, \mathbf{g}_k} \begin{bmatrix} \nabla_{\mathbf{x}_k} \text{SINR} \\ \nabla_{\mathbf{g}_k} \text{SINR} \end{bmatrix}^T$$

$$= \begin{bmatrix} H_{\mathbf{x}_k} \text{SINR} & \nabla_{\mathbf{x}_k} (\nabla_{\mathbf{g}_k} \text{SINR})^T \\ \nabla_{\mathbf{g}_k} (\nabla_{\mathbf{x}_k} \text{SINR})^T & H_{\mathbf{g}_k} \text{SINR} \end{bmatrix}.$$

For the diagonal blocks we have (3.15) and

$$H_{\mathbf{g}_k} \text{SINR} = \mathbf{w}_1 (2\mathbf{K}\mathbf{g}_k)^T - \mathbf{w}_2((\mathbf{C}+\mathbf{C}^T)\mathbf{g}_k)^T$$
$$+ \frac{1}{\mathbf{g}_k^T \mathbf{C} \mathbf{g}_k} 2\mathbf{K} - \frac{\mathbf{g}_k^T \mathbf{K} \mathbf{g}_k}{\left(\mathbf{g}_k^T \mathbf{C} \mathbf{g}_k\right)^2}(\mathbf{C}+\mathbf{C}^T),$$

where

$$\mathbf{w}_1 := \frac{-1}{(\mathbf{g}_k^T \mathbf{C} \mathbf{g}_k)^2}(\mathbf{C}+\mathbf{C}^T)\mathbf{g}_k$$

$$\mathbf{w}_2 := \frac{1}{(\mathbf{g}_k^T \mathbf{C} \mathbf{g}_k)^2} 2\mathbf{K}\mathbf{g}_k - \frac{2\mathbf{g}_k^T \mathbf{K} \mathbf{g}_k}{\left(\mathbf{g}_k^T \mathbf{C} \mathbf{g}_k\right)^3}(\mathbf{C}+\mathbf{C}^T)\mathbf{g}_k.$$

The off-diagonal blocks can be written as

$$\nabla_{\mathbf{g}_k}(\nabla_{\mathbf{x}_k}\text{SINR})^T = \mathbf{u}_1(\nabla_{\mathbf{x}_k}s)^T - \mathbf{u}_2(\nabla_{\mathbf{x}_k}n)^T - \mathbf{u}_3(\nabla_{\mathbf{x}_k}c)^T$$
$$+ \frac{2s}{n \cdot c}\nabla_{\mathbf{g}_k}(\nabla_{\mathbf{x}_k}s)^T - \frac{s^2}{n^2 c}\nabla_{\mathbf{g}_k}(\nabla_{\mathbf{x}_k}n)^T$$
$$= \left(\nabla_{\mathbf{x}_k}(\nabla_{\mathbf{g}_k}\text{SINR})^T\right)^T,$$

where

$$\mathbf{u}_1 := \nabla_{\mathbf{g}_k}\frac{2s}{n \cdot c} = \frac{2}{n \cdot c}\nabla_{\mathbf{g}_k}s - \frac{2s}{n^2 c}\nabla_{\mathbf{g}_k}n$$

$$\mathbf{u}_2 := \nabla_{\mathbf{g}_k}\frac{s^2}{n^2 c} = \frac{2s}{n^2 c}\nabla_{\mathbf{g}_k}s - \frac{2s^2}{n^3 c}\nabla_{\mathbf{g}_k}n$$

$$\mathbf{u}_3 := \nabla_{\mathbf{g}_k}\frac{s^2}{nc^2} = \frac{2s}{nc^2}\nabla_{\mathbf{g}_k}s - \frac{s^2}{n^2 c^2}\nabla_{\mathbf{g}_k}n.$$

With $\mathbf{Q} := \text{diag}(\mathbf{H}\mathbf{V}_1\mathbf{x}_k)$ follows

$$\nabla_{\mathbf{g}_k}s = \mathbf{H}\mathbf{V}_1\mathbf{x}_k \odot \mathbf{H}\mathbf{V}_1\mathbf{x}_k$$
$$\nabla_{\mathbf{g}_k}n = 8\left(\mathbf{H}\mathbf{V}_1\mathbf{x}_k\mathbf{x}_k^T\mathbf{V}_1^T\mathbf{H}^T \odot \boldsymbol{\Sigma}_{nn} + \mathbf{x}_k^T\mathbf{x}_k\left(\boldsymbol{\Sigma}_{nn} \odot \boldsymbol{\Sigma}_{nn} - \boldsymbol{\Sigma}_{n'n} \odot \boldsymbol{\Sigma}_{n'n}\right)\right)\mathbf{g}_k$$

and

$$\nabla_{\mathbf{g}_k}(\nabla_{\mathbf{x}_k}s)^T = 2\mathbf{Q}\mathbf{H}$$
$$\nabla_{\mathbf{g}_k}(\nabla_{\mathbf{x}_k}n)^T = 16\left(\mathbf{Q}\boldsymbol{\Sigma}_{nn}\mathbf{G}\mathbf{H} + (\boldsymbol{\Sigma}_{nn} \odot \boldsymbol{\Sigma}_{nn} - \boldsymbol{\Sigma}_{n'n} \odot \boldsymbol{\Sigma}_{n'n})\mathbf{g}_k\mathbf{x}_k^T\mathbf{V}_1^T\right).$$

After a sufficient number of iterations K, the solution can finally be obtained by

$$\begin{bmatrix} \tilde{\mathbf{p}}_K \\ \mathbf{g}_K \end{bmatrix} = \begin{bmatrix} \mathbf{V}_1 : \mathbf{I}_N \end{bmatrix} \cdot \begin{bmatrix} \mathbf{x}_K \\ \mathbf{g}_K \end{bmatrix}$$

Numerical experiments shows good convergence behavior for the Newton's Method for different channels and random starting vectors. For all realization the same maximum was found. Therefore, we conjecture that the presented algorithm solves the optimization problem.

3.4.3 Statistical channel knowledge

To avoid channel estimation at the low complexity receiver, we propose to identify the multipath conditions from the location of the SN. However, localization of SNs is always subject to estimation errors and only possible up to a certain accuracy. This requires robust optimization algorithms that do not depend on an individual channel realization but rather on the statistics of the channel conditions at a certain position. Hence, we model the channel impulse response as a nonstationary random process as shown in Chapter 2, cf. (2.1), and derive the optimization algorithms in the following.

With the additional assumption that the channel impulse response is Gaussian distributed, this results in an optimization based on the mean and autocorrelation function of the channel impulse response. In this case, we model $\begin{bmatrix} \tilde{h}(T), \dots, \tilde{h}(NT) \end{bmatrix}^T$ as jointly Gaussian distributed random vector with mean $\boldsymbol{\mu}_h$ and covariance matrix $\boldsymbol{\Sigma}_h$. In practice, knowledge of the parameters can for example be obtained by measurements with a higher complexity node or based on assumptions of a certain propagation environment or location of the nodes.

First, we present an optimization algorithm, which approximates the minimum mean bit error rate (mBER) based on a Gauss approximation of the energy detector output. Second, optimization of the mean SINR (mSINR) is derived, where expectation of numerator and denominator are taken separately. Moreover, this leads to a low SINR approximation (mLSA) for the transmitter optimization.

Minimization of mean BER (mBER): Since the SINR is now a random variable, it is not anymore reasonable to maximize its value. Instead, we minimize the mean bit error probability P_e to achieve the best performance on average, i.e.

$$\mathbf{p}_{\text{mBER}}^* = \underset{\mathbf{p} \in \mathbb{R}^N : \|\mathbf{p}\| = 1}{\arg\min} \; \mathsf{E}_h \left[P_e \right] \tag{3.16}$$

35

or

$$g^*_{\text{mBER}} = \arg\min_{g \in \mathbb{R}^N} E_h\left[P_e\right], \tag{3.17}$$

for the transmitter or post-detection filter, respectively. The expectation is taken with respect to the channel impulse response $\left[\tilde{h}(T), \ldots, \tilde{h}(NT)\right]^T$. However, due to the nonlinear receiver processing, a closed form expression for P_e is difficult to obtain[2]. In order to keep the problem tractable, we approximate the energy detector output as Gaussian distributed. The validity of this assumption for a similar problem has been investigated in [73]. With the Gaussian assumption, the bit error probability P_e simplifies to

$$P_e \approx Q\left(\sqrt{\text{SINR}}\right).$$

To compute the mean of P_e over the channel realizations, we propose to apply Monte Carlo Integration:

$$E_h\left[P_e\right] \approx \frac{1}{M_h}\sum_{i=1}^{M_h} Q\left(\sqrt{\text{SINR}_i}\right), \tag{3.18}$$

where SINR_i is computed with the ith realization of M_h randomly drawn channels according to the distribution of the channel.

We use again Newton's method to solve the optimization problem. The gradient of (3.18) is given by

$$\nabla_{\boldsymbol{\Omega}} E_h\left[P_e\right] \approx \frac{1}{M_h}\sum_{i=1}^{M_h} -\frac{e^{-\frac{\text{SINR}_i}{2}}}{\sqrt{8\pi\text{SINR}_i}}\nabla_{\boldsymbol{\Omega}}\text{SINR}_i,$$

where $\boldsymbol{\Omega}$ is either $\mathbf{x}$ or $\mathbf{g}$. The Hessian matrix computes to

$$H_{\boldsymbol{\Omega}} E_h\left[P_e\right] \approx \frac{1}{M_h}\sum_{i=1}^{M_h} \xi_i \cdot H_{\boldsymbol{\Omega}}\,\text{SINR}_i + \psi_i \cdot \nabla_{\boldsymbol{\Omega}}\text{SINR}_i\left(\nabla_{\boldsymbol{\Omega}}\text{SINR}_i\right)^T,$$

where

$$\xi_i = -\frac{1}{2\sqrt{2\pi\text{SINR}_i}} \cdot e^{-\frac{\text{SINR}_i}{2}}$$

---

[2]The PDF of the energy detector output has been analyzed in detail based on measurements and channel models in [25].

and

$$\psi_i = \frac{1}{4\sqrt{2\pi}}(\text{SINR}_i^{-\frac{3}{2}} + \text{SINR}_i^{-\frac{1}{2}}) \cdot e^{-\frac{\text{SINR}_i}{2}}.$$

Since $\mathsf{E}_h[P_e]$ may have many local minimums, we cannot guarantee to find the optimal solution. However, in the following we will describe two approximations that can serve as starting vectors for the iterative solution to the optimization problem.

Maximization of mean SINR *(mSINR)*: As an alternative to (3.16), the optimization can be performed with respect to the mean signal and mean interference plus noise power, i.e. for the transmitter

$$
\begin{aligned}
\mathbf{P}_{\text{mSINR}}^* &= \underset{\tilde{\mathbf{p}} \in \mathbb{R}^N}{\arg\max} \frac{\mathsf{E}_h\left[\tilde{\mathbf{p}}^T \mathbf{A} \tilde{\mathbf{p}}\right]^2}{\mathsf{E}_h\left[\tilde{\mathbf{p}}^T \mathbf{B} \tilde{\mathbf{p}} \cdot \tilde{\mathbf{p}}^T \tilde{\mathbf{p}}\right]} \\
&= \underset{\tilde{\mathbf{p}} \in \mathbb{R}^N}{\arg\max} \frac{\left(\tilde{\mathbf{p}}^T \bar{\mathbf{A}} \tilde{\mathbf{p}}\right)^2}{\tilde{\mathbf{p}}^T \bar{\mathbf{B}} \tilde{\mathbf{p}} \cdot \tilde{\mathbf{p}}^T \tilde{\mathbf{p}}}.
\end{aligned}
\tag{3.19}
$$

where $\bar{\mathbf{A}} := \mathsf{E}_h[\mathbf{A}]$ and $\bar{\mathbf{B}} := \mathsf{E}_h[\mathbf{B}]$. Note that the expectation operations in the numerator and denominator are taken independently. Moreover, for the signal term in the numerator, the square of the mean is considered instead of the second moment. This is done to simplify the problem and corresponds to a lower bound of the mean signal power, since

$$\mathsf{E}_h\left[\tilde{\mathbf{p}}^T \mathbf{A} \tilde{\mathbf{p}}\right]^2 \le \mathsf{E}_h\left[(\tilde{\mathbf{p}}^T \mathbf{A} \tilde{\mathbf{p}})^2\right].$$

Using (3.6), we obtain for the matrix in the numerator

$$\bar{\mathbf{A}} = \sum_{k=1}^{N} g_k \mathsf{E}_h\left[\mathbf{h}_k \mathbf{h}_k^T\right], \tag{3.20}$$

where $\mathbf{h}_k^T$ denotes the k-th row of $\mathbf{H}$ and g_k the k-th element of $\mathbf{g}$. The matrix $\mathsf{E}_h\left[\mathbf{h}_k \mathbf{h}_k^T\right]$ depends on the mean and covariance of the channel impulse response taps. The (i,j)-th element of the matrix in the denominator computes to

$$\left[\bar{\mathbf{B}}\right]_{i,j} = 4\mathsf{E}_h\left[\mathbf{f}_i^T \mathbf{G} \mathbf{\Sigma}_{nn} \mathbf{G}^T \mathbf{f}_j\right] + \begin{cases} \sigma_\gamma^2 & \text{for } i = j \\ 0 & \text{else,} \end{cases}$$

where $\mathbf{f}_i$ is the ith column of $\mathbf{H}$. The expectation of the quadratic form yields

$$\mathsf{E}_h\left[\mathbf{f}_i^T \mathbf{G}\boldsymbol{\Sigma}_{nn}\mathbf{G}^T\mathbf{f}_j\right] = \mathrm{Tr}\left(\mathbf{G}\boldsymbol{\Sigma}_{nn}\mathbf{G}^T\mathsf{E}_h\left[\mathbf{f}_j\mathbf{f}_i^T\right]\right).$$

The optimization problem (3.19) is of the same form as (3.10). Compared to the case of full channel knowledge, only the values of the matrices in the numerator and denominator change. Hence, the same optimization algorithm can be used to find the optimized pulse shape.

For the post-detection filter, the maximization of the mean SINR can be written as

$$
\begin{aligned}
\mathbf{g}^*_{\text{mSINR}} &= \underset{\mathbf{g}\in\mathbb{R}^N}{\arg\max}\ \frac{\mathsf{E}_h\left[\mathbf{g}^T\mathbf{K}\mathbf{g}\right]}{\mathsf{E}_h\left[\mathbf{g}^T\mathbf{C}\mathbf{g}\right]} \\
&= \mathbf{v}_{\max}\left\{\mathsf{E}_h\left[\mathbf{K}\right],\mathsf{E}_h\left[\tfrac{1}{2}(\mathbf{C}+\mathbf{C}^T)\right]\right\}.
\end{aligned}
\tag{3.21}
$$

For Gaussian distributed channel taps, the (i,j)-th element of $\bar{\mathbf{K}} = \mathsf{E}_h\left[\mathbf{K}\right]$ is given by

$$\left[\bar{\mathbf{K}}\right]_{i,j} = \mathbf{p}^T\mathsf{E}_h\left[\mathbf{h}_i\mathbf{h}_i^T\right]\mathbf{p}\cdot\mathbf{p}^T\mathsf{E}_h\left[\mathbf{h}_j\mathbf{h}_j^T\right]\mathbf{p} + 2\mathbf{p}^T\mathsf{E}_h\left[\mathbf{h}_i\mathbf{h}_j^T\right]\mathbf{p}\cdot\mathbf{p}^T\mathsf{E}_h\left[\mathbf{h}_i\mathbf{h}_j^T\right]\mathbf{p},$$

where $\mathbf{h}_i^T$ is the ith row of $\mathbf{H}$. The expected values of $\mathbf{h}_i\mathbf{h}_j^T$ again depend only on the mean and covariance of the channel impulse response. Likewise, the (i,j)-th element of $\bar{\mathbf{C}} = \mathsf{E}_h\left[\mathbf{C}\right]$ can be written as

$$\left[\bar{\mathbf{C}}\right]_{i,j} = 4\left(\mathbf{p}^T\mathsf{E}_h\left[\mathbf{h}_i\mathbf{h}_j^T\right]\mathbf{p}\right)\left[\boldsymbol{\Sigma}_{nn}\right]_{i,j} + 4\left[\boldsymbol{\Sigma}_{nn}\right]_{i,j}^2 - 4\left[\boldsymbol{\Sigma}_{n'n}\right]_{i,j}^2.$$

Maximization of mean Low-SINR Approximation (mLSA): Analogous to the case considered in the previous subsection for full channel knowledge, we perform a low-SINR approximation by neglecting the mixed signal, noise and interference term. This leads to the simplified optimization problem for the transmitter:

$$
\begin{aligned}
\mathbf{p}^*_{\text{mLSA}} &= \underset{\tilde{\mathbf{p}}\in\mathbb{R}^N}{\arg\max}\ \frac{\tilde{\mathbf{p}}^T\bar{\mathbf{A}}\tilde{\mathbf{p}}}{\tilde{\mathbf{p}}^T\tilde{\mathbf{p}}} \\
&= \mathbf{v}_{\max}\left\{\bar{\mathbf{A}}\right\}
\end{aligned}
\tag{3.22}
$$

The denominator is then independent of the pulse shape and does not need to be considered for the optimization. Thus, mLSA corresponds to finding the pulse shape that maximizes the mean of the energy detector output (or, more precisely, the mean of the decision variable z_0) with respect to noise, interference and channel. The optimization problem is solved

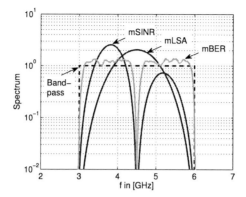

Fig. 3.5: Spectrum of transmit pulse from different optimization methods with statistical channel knowledge and narrowband interference.

by the principle eigenvector of $\bar{\mathbf{A}}$. To compute the matrix $\bar{\mathbf{A}}$, only the knowledge of the channel covariance matrix and mean and the post-detection filter is necessary. This makes this approximation a promising transmission scheme which requires only limited channel information at the transmitter. Moreover, the solution $\mathbf{p}^*_{\mathrm{mLSA}}$ serves as a good initial value for iterative solution to (3.19).

Fig. 3.5 shows the result of a transmitter optimization to compare the three optimization methods mBER, mSINR and mLSA. The spectrum of the transmit pulses is plotted, optimized with statistical channel knowledge and the presence of an interference source at 4.5 GHz. The mBER optimization is based on $M_h = 10^5$ randomly drawn channel realizations. For this example a setting has been chosen[3], which shows the characteristics of the different methods. It can be seen that mBER as well as mSINR place a notch in the spectrum at the frequency of the interferer. The reason for this is to minimize the mixed interference plus noise term β. However, this is not the case for mLSA, since here the mixed term is neglected. The mBER pulse seems to approximate the bandpass pulse and uses almost the whole transmission band with the exception of the interference frequency.

---

[3]Optimization parameters: $B = 3\,\mathrm{GHz}$, $f_c = 4.5\,\mathrm{GHz}$, SIR $= 3\,\mathrm{dB}$, Gaussian channel taps, exponential PDP with $\sigma_h = 10\,\mathrm{ns}$, Post-detection filter: First-order low-pass with $f_{\mathrm{cutoff}} = 25\,\mathrm{MHz}$ (as defined in Section 3.5).

3.5 Performance Evaluation

This section evaluates the performance of the transmitter and receiver optimization. First, we evaluate the performance gains that can be achieved if the channel is assumed to be known at the transmitter or receiver. For this analysis we use a channel model. To consider a real environment, in a second step the performance of the proposed optimization schemes is evaluated based on measured channels. The performance gains of transmitter and receiver optimization based on statistical channel knowledge are quantified and it is shown that interference can effectively be suppressed on the receiver as well as transmitter. Before we discuss the results, we define the system parameters and describe the simulation setup.

For the performance evaluation we consider a UWB system with a bandwidth of $B = 3\,\text{GHz}$ at a center frequency of $f_c = 4.5\,\text{GHz}$. The duration of the PPM half slot is chosen as $T_{\text{ppm}} = 50\,\text{ns}$. With a duty cycle of 100%, this would correspond to a data rate of 10 Mbps. As figure of merit we consider the BER versus signal-to-noise-ratio, which is defined as E_b/N_0, with E_b denoting the energy per bit and $N_0/2$ the noise power spectral density. We consider a single narrowband interferer with bandwidth $\text{BW}_\text{i} = 10\,\text{MHz}$ at $f_\text{i} = 4.5\,\text{GHz}$. Simulations show that there is no significant influence of the value of f_i. The power of the interference is measured by the signal-to-interference ratio, which is defined as

$$\text{SIR} = \frac{P_S}{P_I} = \frac{E_b/T_{\text{symb}}}{I_0 \cdot \text{BW}_\text{i}},$$

where $I_0/2$ is the power spectral density of the interference in the considered band. The bandpass filter at the receiver input is assumed to be perfectly band limiting for the considered transmission band from 3 to 6 GHz. Hence, the (i,j)-th element of the noise plus interference covariance matrix is given by

$$[\mathbf{\Sigma}_{nn}]_{i,j} = N_0 \tfrac{2B}{f_s} \operatorname{sinc}\left(\tfrac{B(i-j)}{f_s}\right) \cos\left(2\pi \tfrac{f_c}{f_s}(i-j)\right)$$
$$+ I_0 \tfrac{2\text{BW}_\text{i}}{f_s} \operatorname{sinc}\left(\tfrac{\text{BW}_\text{i}(i-j)}{f_s}\right) \cos\left(2\pi \tfrac{f_\text{i}}{f_s}(i-j)\right),$$

with the simulation sampling frequency $f_s = 30\,\text{GHz}$.

3.5.1 Simulation results based on channel model

To evaluate the performance of the proposed optimization schemes in a generic setting, we consider a simple channel model. This channel model assumes that the bandpass filtered channel impulse responses $\tilde{\mathbf{h}}$ are drawn from a Gaussian distribution with zero mean and covariance matrix $\boldsymbol{\Sigma}_h$, i.e.

$$\tilde{\mathbf{h}} \sim \mathcal{N}(\mathbf{0}, \boldsymbol{\Sigma}_h).$$

The channel impulse responses have an exponential PDP with root mean square (RMS) delay spread $\sigma_h = 10\,\text{ns}$. The channel covariance matrix is given by

$$\boldsymbol{\Sigma}_h = \mathbf{F}_{\text{BP}} \mathbf{D}_{\text{pdp}} \mathbf{F}_{\text{BP}}^T$$

with

$$[\mathbf{F}_{\text{BP}}]_{i,j} = \frac{2B}{f_s} \operatorname{sinc}\left(\frac{B(i-j)}{f_s}\right) \cdot \cos\left(2\pi \frac{f_c}{f_s}(i-j)\right)$$

and the diagonal PDP matrix

$$\left[\mathbf{D}_{\text{pdp}}\right]_{i,i} = \frac{1}{\sigma_h} \exp\left(-\frac{i}{f_s \sigma_h}\right).$$

Based on these randomly generated channel impulse responses, bit error rate simulations with transmitter and receiver optimization have been performed.

Transmitter Optimization: For the transmitter optimization, the post-detection filter has to be fixed and can be chosen arbitrarily. To account for stringent complexity requirements we consider a first-order low-pass filter at the receiver. Thus, for transmitter optimization, the impulse response of the post-detection filter is given by

$$g(kT) = g_0 \exp\left(-\frac{k 2\pi f_{\text{cutoff}}}{f_s}\right), \quad k > 0.$$

The cutoff frequency is set to $f_{\text{cutoff}} = 25\,\text{MHz}$.

An example for a channel impulse response and corresponding pulses is shown in Fig. 3.6. Here, we chose the signal-to-noise-ratio to $E_b/N_0 = 18\,\text{dB}$ and SIR $= 3\,\text{dB}$. Unfortunately, it is difficult to give an intuitive or qualitative description of the optimization outcomes. This is a consequence of the non-linear receiver processing and the structure of the problem at hand. The shape of the optimized pulse in time domain does not give any further insights. However, we observe in the spectrum of the optimized pulses some similarities to the transfer function of the channel, see Fig. 3.7. The results show that the transmitter prefers frequencies where the channel conditions are good. Moreover, the pulses are formed in a way that they suppress

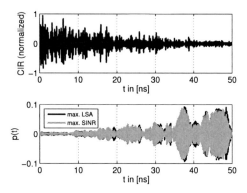

Fig. 3.6: Example of channel impulse response (top) and optimized pulses (bottom)

the narrowband interference and that they are matched to the post-detection filter.

Fig. 3.8 shows the mean BER of the transmitter optimization with full channel knowledge for the fixed receiver, averaged over 100 channel realizations. For comparison, the plot depicts the mean BER with an ideal bandpass pulse with perfect synchronization. The dashed lines show the mean BER of the low-SINR approximation (LSA). Markers label different interference levels, i.e. circles, diamonds and squares correspond to the case without interference, $SIR = 3\,dB$, and $SIR = 0\,dB$, respectively. The plot shows that without interference, more than 6 dB gain can be achieved by the optimized transmitter compared to transmission of an ideal bandpass pulse. Moreover, the precoding can effectively suppress the interference. Whereas for $SIR = 0\,dB$ the BER saturates at about 0.1 for the bandpass pulse, the optimized transmission scheme still achieves an acceptable BER of less than 10^{-3}.

To compare the SINR maximization with LSA, Fig. 3.9 shows in the upper plot the mean SINR versus E_b/N_0 for both schemes. The solid and the dashed lines correspond to maximization of SINR and LSA, respectively. It can be seen that the analytical result (lines) according to (3.3) and the empirically estimated SINR from the simulation (markers) coincide. In low SINR the approximation (LSA) is tight and even in higher SINR the deviation is not large. The lower plot in Fig. 3.9 shows the CDF of the SINR for $E_b/N_0 = 15\,dB$. The dashed line corresponds to LSA and is constantly worse than the SINR maximization.

Receiver Optimization: Fig. 3.10 shows the performance of the receiver optimization with full channel knowledge for a fixed transmit pulse. Here, just an ideal bandpass pulse has been chosen for transmission. The post-detection filter has been optimized for 4000

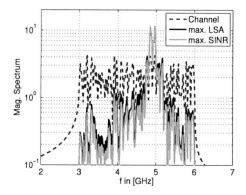

Fig. 3.7: Magnitude of channel transfer function and spectrum of optimized pulses

channel realizations each. We observe that the optimized $g(t)$ shows similarities to the time-reversed and squared channel impulse response, see. Fig. 3.11. This would be optimal under a Gaussian assumption of the energy detector output (channel matched filter). However, the optimized filter can also suppress narrowband interference. The plot shows for comparison the mean BER of a conventional energy detector (ED) with rectangular integration window (dashed line). The length of the integration window is 50 ns. Moreover, the performance is plotted, if a coherent detector would have been used. The coherent receiver is about 3 dB better than the optimized generalized energy detector. The conventional energy detector suffers strongly from narrowband interference. With SIR $= 3$ dB the BER is greater than 10 %. The optimization of the post-detection filter achieves small degradation even in presence of strong narrowband interference.

Joint Optimization: Finally, Fig. 3.12 shows the performance of joint optimization. Additionally, the performance of transmitter-only and receiver-only optimization is plotted for comparison. We observe about 10 dB performance gain compared to bandpass transmission and first-order low-pass filter detection. Transmitter-only and receiver-only optimization lies in between. However, please note that the first-order low-pass filter and the bandpass transmission are arbitrary choices, which influence the performance significantly. The joint optimization can serve as performance bound for a specific implementation. In particular, it relates the performance of a specific choice of post-detection filter or transmit pulse to the optimum. For instance, it can be seen that the choice of a first-order low-pass for transmitter optimization leads to a loss of only about 3 dB compared to the optimum post-detection fil-

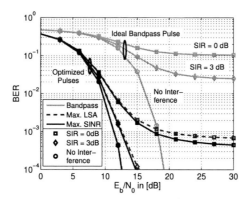

Fig. 3.8: Performance of transmitter optimization with full channel knowledge

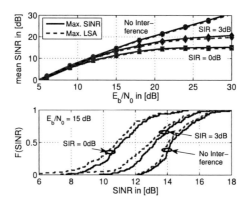

Fig. 3.9: Transmitter optimization: Mean SINR vs. E_b/N_0 and CDF of SINR

ter. This is surprisingly small, when considering the significant difference in implementation complexity. In terms of practicality, the joint optimization is of limited use. It requires full channel state information at the transmitter and receiver and both need to be adaptable to the channel conditions. In this case, it may be favorable to implement coherent detection. Moreover, the jointly optimized transmit pulses use only a single frequency for transmission, see Fig. 3.13. The upper plot depicts the magnitude spectrum of the pulse from joint optimization for different signal-to-noise ratios E_b/N_0. The transmitter does not use the full UWB spectrum but rather degrades to narrowband transmission on the best available frequency.

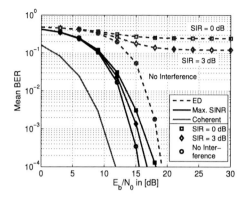

Fig. 3.10: Performance of receiver optimization with full channel knowledge

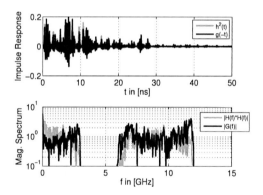

Fig. 3.11: Optimized post-detection filter and squared channel impulse response

The lower plot in Fig 3.13 shows the corresponding post-detection filters.

3.5.2 Measured Channels

To analyze the performance of the proposed optimization schemes in a real environment, an extensive measurement campaign has been performed. The measurements took place in an indoor rich multipath environment. Fig. 3.14 shows the floor plan of the room with 22 regions. For each of these regions of size $27\,\text{cm} \times 56\,\text{cm}$, about 600 channel impulse responses

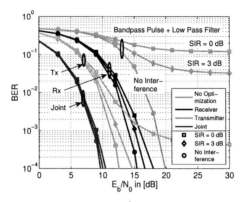

Fig. 3.12: Comparison of joint optimization and transmitter or receiver optimization

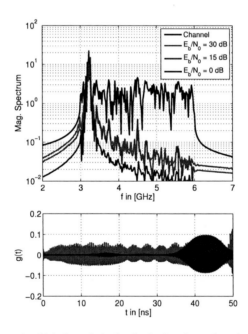

Fig. 3.13: Example of jointly optimized pulse (top) and post-detection filter (bottom)

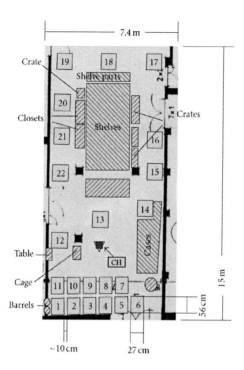

Fig. 3.14: Floor plan of measurement scenario

have been measured. The measurement setup is described in detail in [39, 74]. With these measurements, the performance of location-aware transmitter and receiver optimization is evaluated for each region. We consider the transmission from the CH to low complexity SNs located in different regions. In the first scenario, the receiver is assumed to be fixed and a region specific pulse shape has been determined. In the second scenario, the low complexity SNs may be able to adjust the post-detection filter. The BER performance for both transmission schemes is determined under the influence of narrowband interference. We present the results for two characteristic regions, namely region 9, which is a typical line-of-sight (LOS) situation, and region 17, which is a worst case non-line-of-sight (NLOS) situation.

Transmitter optimization: Fig. 3.15 and 3.16 show the mean BER for the first scenario for region 9 and region 17, respectively. The dashed lines show the BER without interference and the solid lines the case of $SIR = 3\,dB$. As a reference, the performance of transmission

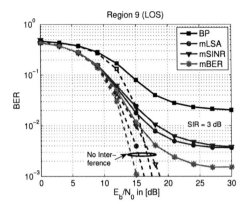

Fig. 3.15: Performance of transmitter optimization with statistical channel knowledge for a LOS region

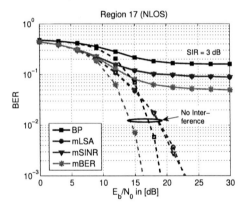

Fig. 3.16: Performance of transmitter optimization with statistical channel knowledge for a NLOS region

of an ideal bandpass pulse is plotted (squares). In the LOS situation, the performance is better than in NLOS, because the direct path contains the most energy. Additionally, the channel impulse responses from these regions have more structure, which enables the transmitter optimization to work more effectively. For the case without interference, the mBER pulse (asterisks) is about 2.5 dB better than the bandpass pulse. Narrowband interference can effectively be suppressed leading to 7.5 to 31 % BER reduction for mBER at high SNR

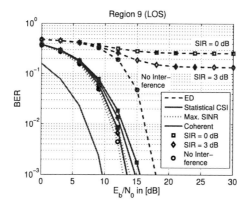

Fig. 3.17: Performance of receiver optimization with statistical channel knowledge for a LOS region

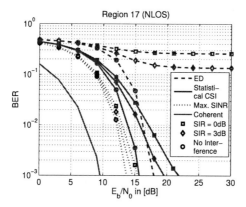

Fig. 3.18: Performance of receiver optimization with statistical channel knowledge for a NLOS region

for LOS and NLOS, respectively. The mSINR (triangles) and mLSA (circles) optimizations lie in between if E_b/N_0 is not too high. Interestingly, for the LOS region, the low-SINR approximation (mLSA) performs better than mSINR. However, for the LOS region the approximations still show performance gains. For the NLOS case without interference, it can be observed that the approximations perform worse than the simple transmission of a bandpass pulse. This is because especially at high E_b/N_0 the maximization of mean SINR does not minimize the bit error probability. In this case, some bad channel realizations dominate the BER performance, which is not considered by the mean SINR objective function.

Receiver Optimization: Fig. 3.17 and 3.18 show the receive filter optimization for region 9 and region 17, respectively. Again, at the transmitter an ideal bandpass pulse is chosen. The plots show the post-detection filter optimization based on the region knowledge (solid lines) as well as with full channel knowledge (dotted lines). Surprisingly, the optimization based on the statistical channel knowledge almost achieves the performance with full channel knowledge and the interference is successfully suppressed. For the NLOS case (Fig. 3.18), the performance is worse than for the LOS case. The reason for this is that the channel in this regions differ more and have less structure than in the LOS region. However, it can be seen that the regional channel knowledge enables good detection performance and makes the receiver less vulnerable to narrowband interference compared to conventional energy detection. Moreover, it is robust and only the region where the receiver is located must be known to adapt the post-detection filter.

3.6 Implementation Issues and Complexity Analysis

Noncoherent receivers are the key to low cost and low complexity implementations of UWB sensor nodes. Coherent detectors provide a better BER performance but require either very high sampling rates or a high number of rake fingers, which make them impractical for low-power devices. Implementations of noncoherent UWB receivers with energy detection have been shown with a power consumption as low as 136 mW at 5 Mbps data rate (e.g. [75]) or, when applying a low-duty cycle operation, even less than 1 mW (see [58]) for 500 kbps. In this chapter, we present two efficient ways to partially recover the performance loss of energy detection receivers. Both come with little implementation overhead, which is described in the following.

- *Transmitter optimization:* The precoding approach shifts all complexity from the receiver to the transmitter at the CH. Hence, the receiver is not required to adapt to the channel conditions and a fixed post-detection filter can be exploited, e.g. a simple first order low-pass filter. To apply the transmitter optimization, the CH must be able to adapt to the transmit signal. Although this comes with a higher complexity, it is a reasonable assumption since the CH is a full-function device with less stringent constraints on power and cost. Moreover, it is not necessary to perform a convolution with the optimized pulses at the transmitter. The adaptation can be realized with a look-up table with the predetermined pulses. The optimization is based on the location of the SN. Thus, it can be avoided to estimate the channel at the SNs, which would be impractical due to their non-coherent receivers. Instead of the complete channel impulse response, only the region of the SN is needed. The narrowband interference suppression is based on the covariance matrix of the interference. This can be obtained either by a priori knowledge of the coexisting narrowband communication systems or from measurements at the CH. The latter assumes that the CH is exposed to the same interference conditions as the SN, which is reasonable for short range systems.

- *Receiver optimization:* For the implementation of the receiver optimization, the SNs require an adaptable post-detection filter. For an analog implementation, the optimized impulse response can serve as design guideline for the filter and additionally it can be used as a benchmark for implementation. Even a digital implementation of the post-detection filter is favorable in terms of complexity compared to coherent detection. To trade performance for a lower complexity, the post-detection filter can be optimized in a subspace, i.e. using the substitution $g \mapsto W\tilde{g}$, where the columns of W are the basis vectors of the considered subspace. This enables to restrict the optimization to specific filter structures or to limit the bandwidth or sampling frequency. For location-aware adaption, the SN does not need to estimate the channel. Its location or its region can simply be estimated by the CH and then send to the SN. This saves substantial channel estimation and dissemination overhead.

Location-aware communication requires a database with the mapping of locations to channel statistics. For a stationary environment (as typical for indoor scenarios), this can be build up in a calibration phase from training data. Due to the statistical channel model, the optimization is robust against small changes of the propagation environment. The performance of the precoding depends on the size of the region and on the characteristics of the multipath. A smaller region, and thus better channel knowledge, would lead to better performances. The receiver optimization is more robust to channel uncertainty.

3.7 Conclusion

Location information can help to improve the detection performance of low complexity UWB communication. We studied location-aware transmitter and receiver optimizations for generalized energy detection receivers with binary pulse position modulation. The performance evaluation based on measurements shows a 2-5 dB gain compared to simple energy detection. This provides a promising synergy of localization and communication for low complexity UWB sensor networks.

Chapter 4

Multiuser Precoding

In this chapter[1], we focus on the downlink of the UWB sensor network, i.e. communication from a cluster head to the sensor nodes. Due to the limited complexity and the generalized energy detection receiver, the data rate is limited by the delay spread of the channel. To overcome this problem, we propose a novel precoding scheme to transmit to several nodes simultaneously. This way, the sum data rate can be increased, while low complexity of sensor nodes is maintained.

This chapter is structured as follows. In Section 4.2, the system model is described. Section 4.3 introduces the considered SINR expression, which is optimized in Section 4.4. The performance evaluation is presented in Section 4.5 and conclusions are drawn in Section 4.6.

4.1 Dense Networks: Low data rate despite Gigahertz bandwidth

With increasing data rate, non-coherent detection of UWB signals becomes difficult due to inter-symbol interference (ISI) from multipath. For very low-complexity energy detection receivers, the symbol timing cannot be much shorter than the excess delay of the strong multipath components. Otherwise, the multipath perturbs the next symbol and makes it difficult for the receiver to decode. Thus, the data rate of low complexity UWB nodes is limited by intersymbol interference, which leads to a paradox situation: Only low data rates can be achieved, even though several Gigahertz of bandwidth are available. For dense sensor networks with numerous nodes, this may result in severe restrictions on the per node throughput.

---

[1]Parts of this chapter have been published in [44].

Let us for example consider a multipath channel with a delay spread of 25 ns. With BPPM and an energy detection receiver, this would result in a maximum data rate of 20 Mbps. Whereas this is more than enough for a single sensor node, it becomes a problem when the network grows large. With millions of nodes, only a very low per node data rate can be achieved because each node is served rarely. This is a strange situation because a very high bandwidth is available for the data transmission. Even with the simplest form of binary pulse amplitude modulation with Nyquist pulses, the UWB channel from 3.1 to 10.6 GHz would support a data rate of up to 7.5 Gbps. In contrast to conventional communication systems, it is not the bandwidth that limits the data rate of UWB sensor networks. It is rather the time dispersive channel and the hardware complexity that dictate the throughput of the network.

To handle ISI is one of the standard problems in wireless communications and the state-of-the-art offers numerous powerful solutions for many applications. The optimal coherent detector resolves the ISI by considering all different combinations of symbols and chooses the most probable. This can be implemented by the Viterbi algorithm. However, the number of states grows exponentially with the data rate (assuming a fixed modulation alphabet). Suboptimal approaches are linear equalizers such as Wiener filters or decision feedback architectures. The receiver complexity can be reduced by an appropriate precoding at the transmitter. Today, probably the most popular approach is OFDM, where the data stream is divided onto many subcarriers. Each of the subcarrier has a low data rate and is thus robust against ISI. The efficient computation of the fast Fourier transform (FFT) makes OFDM favorable for the implementation. However, it still requires a linear receiver with full rate sampling and high precision, which would be far too complex for low complexity UWB sensor networks.

For non-coherent UWB receivers, different approaches for ISI suppression have been presented in literature:

- The optimal non-coherent detectors that account for ISI are presented in [69]. The authors derive the ML receivers for BPPM in presence of ISI for different categories of statistical CSI, such as instantaneous power delay profile (IPDP) or APDP. It can be seen that the optimal decision rules depend on the perfectly sampled receive signals, which is not an option for low complexity sensor nodes.

- MLSE post-detection for energy detection receivers is proposed and analyzed in [76]. Whereas this approach is suited for the generalized energy detection receiver, it would still require extensive digital processing including estimation of the ISI.

- Prominent approaches for transmitter optimization are time reversal [63] or channel phase precoding [66, 77]. When the time reversed channel is used as transmit pulse, the multipath components add up coherently at zero time lag and average out at all others. That way, the intersymbol interference is reduced and the per node data rate can be increased. Channel phase precoding does the same but takes only the sign of the multipath components into account. In [67], time reversal is combined with a pre-equalizer.

Instead of trying to suppress the ISI and to increase the per node data rate, we propose a total different approach. We aim to increase the sum data rate by transmission of several data streams simultaneously. This has significant advantages. The sum data rate can be increased, without additional processing at the low complexity nodes. The receiver remains the generalized energy detector with minimum complexity. For the sensor nodes, the precoding is completely transparent and the processing is exactly the same as for single user transmission. Note that the per node data rate is kept constant. Whereas the throughput can significantly be increased, the low processing speed at the sensor nodes can be maintained. Considering e.g. an FDMA approach, the nodes need to be able to tune to different frequencies, which adds complexity to the receiver. All sensor nodes can have the same receiver for the approach that is presented in this chapter.

To obtain the multiuser precoding pulses, we derive a SINR expression for the pulse position modulated downlink communication with UWB generalized energy detection receivers. Note that in this case the interference is considered to be caused by other nodes, i.e. the interference originates from multiuser transmission. This is in contrast to Chapter 3, where we considered the interference from other narrowband systems, which cannot be influenced. Based on the SINR expression, we formulate the optimization problem to maximize the minimum SINR of all nodes. This is converted to a quasi-convex problem, which can be solved by standard algorithms. Moreover, we extend the precoding optimization to statistical and location-aware channel knowledge. This enables optimization of simultaneous transmission to nodes located in different regions and saves channel estimation overhead. The performance of the proposed precoding scheme is evaluated based on measurements.

4.2 System Model

The system model that we consider in this chapter is shown in Fig 4.1. It is the extension of Fig. 2.2 to downlink multiuser transmission. The transmitter simultaneously sends K streams of pulse position modulated data. The transmitter inputs $b_1(t), \ldots, b_K(t)$ are given

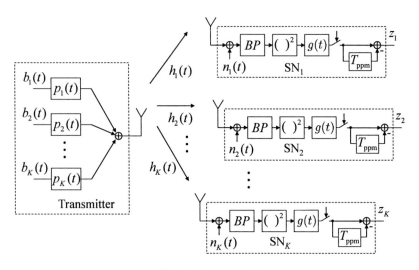

Fig. 4.1: System model

by

$$b_k(t) = \sum_l c_{k,l} \cdot \delta\left(t - a_k[l]T_{\text{ppm}} - lT_{\text{symb}}\right), k = 1, \dots, K.$$

The transmit data of time-slot l, which is intended for node k is denoted by $a_k[l]$. We assume equiprobable $a_k[l] \in \{0,1\}$, which are i.i.d. for all $k = 1, \dots, K$ and time-slots l. Note that the polarity of the pulses is scrambled by random $c_{k,l} \in \{-1,1\}$ to avoid discrete lines in the transmit spectrum. The data streams are linearly precoded by a convolution with $p_1(t), \dots, p_K(t)$, respectively, and then they are added up and transmitted over the same transmit antenna.

The multipath channel to node k is characterized by its impulse response $h_k(t)$. The receive signal is perturbed by independent white Gaussian noise $n_k(t)$ with power spectral density $N_0/2$. Two samples per symbol are taken at the output of the generalized energy detection receivers. They are subtracted and give the decision variables z_k. If z_k is greater than zero, node k decides for a "0" and if it is less than zero, it decides for a "1". In the following, we consider only one time-slot and omit the index l.

4.3 Signal-to-Interference-and-Noise Ratio for Multiuser Transmission

For the derivation of an SINR expression of this setup, we consider the discrete time model with sampling period T and we assume that edge effects are negligible. In vector notation, the receiver output for node k can be written as:

$$z_k = \mathbf{g}^T \left[\left(\sum_{i=1}^{K} a_i \mathbf{H}_k \mathbf{p}_i + \mathbf{n}_k \right) \odot \left(\sum_{j=1}^{K} a_j \mathbf{H}_k \mathbf{p}_j + \mathbf{n}_k \right) \right]$$
$$- \mathbf{g}^T \left[\left(\sum_{i=1}^{K} \bar{a}_i \mathbf{H}_k \mathbf{p}_i + \mathbf{n}_k' \right) \odot \left(\sum_{j=1}^{K} \bar{a}_j \mathbf{H}_k \mathbf{p}_j + \mathbf{n}_k' \right) \right] \tag{4.1}$$

The first line corresponds to the pulses that are transmitted in the first PPM slot and the second line to the second PPM slot, where $\bar{a}_i := 1 - a_i$, $i = 1, \ldots, K$. All vectors are of dimension $N = T_{\text{ppm}}/T$:

- Transmit pulses $\mathbf{p}_i = [p_i(T), \ldots, p_i(NT)]^T$

- Noise of the first and the second half of the time-slot $\mathbf{n}_k, \mathbf{n}_k' \sim \mathcal{N}(0, \Sigma_{nn})$

- Post-detection filter $\mathbf{g} = [g(NT), \ldots, g(T)]^T$, which is stacked into the vector in reverse order

The $(N \times N)$-channel matrix $\mathbf{H}_k$ has Toeplitz structure with the zero-padded and shifted channel impulse response $[\tilde{h}_k(T), \ldots, \tilde{h}_k(NT)]^T$ on its columns. The bandpass is incorporated into the channel impulse response and the covariance matrix of the noise Σ_{nn}. Note that this input/output relation does not account for ISI of the single data streams. It is assumed that the channel excess delay is smaller than T_{ppm}. This is a reasonable assumption for the derivation of the precoding optimization, because we aim to increase the sum data rate by simultaneous transmission to several nodes. This keeps the per node data rate small.

Expanding the element-wise multiplication and rearranging terms of (4.1) yields

$$z_k = \sum_{i=1}^{K} \sum_{j=1}^{K} (a_i a_j - \bar{a}_i \bar{a}_j) \mathbf{p}_i^T \mathbf{H}_k^T \mathbf{G} \mathbf{H}_k \mathbf{p}_j$$
$$+ 2\mathbf{g}^T \left(\sum_{i=1}^{K} (a_i \mathbf{n}_k - \bar{a}_i \mathbf{n}_k') \odot (\mathbf{H}_k \mathbf{p}_i) \right)$$
$$+ \mathbf{g}^T \left(\mathbf{n}_k \odot \mathbf{n}_k - \mathbf{n}_k' \odot \mathbf{n}_k' \right). \tag{4.2}$$

The first line corresponds to the squared signal terms, the second line to the mixed signal and noise terms, and the third line to the squared noise. To keep the problem tractable, we neglect the mixed term for the precoding optimization, which corresponds to a low signal-to-noise ratio (SNR) approximation. The SINR-expression for node k that we use for optimization is defined as

$$\text{SINR}_k := \frac{\mathbf{p}_k^T \mathbf{H}_k^T \mathbf{GH}_k \mathbf{p}_k}{\sum_{i \neq k} \sum_{j \neq k} \sigma_{ij} \mathbf{p}_i^T \mathbf{H}_i^T \mathbf{GH}_k \mathbf{p}_j + \sigma_n^2}, \tag{4.3}$$

where $\sigma_{ij} = 1$ for $i = j$, and $\sigma_{ij} = 1/2$ for $i \neq j$. The numerator contains the desired signal component, whereas the denominator collects the interference terms and the noise term σ_n^2, which is independent of $\mathbf{p}_i$ and proportional to N_0.

Stacking all transmit pulses into one large vector $\bar{\mathbf{p}} = \left[\mathbf{p}_1^T, \ldots, \mathbf{p}_K^T \right]^T$, we obtain

$$\text{SINR}_k = \frac{\bar{\mathbf{p}}^T \left(\mathbf{E}_k \otimes \mathbf{H}_k^T \mathbf{GH}_k \right) \bar{\mathbf{p}}}{\bar{\mathbf{p}}^T \left(\mathbf{F}_k \otimes \mathbf{H}_k^T \mathbf{GH}_k \right) \bar{\mathbf{p}} + \sigma_n^2},$$

where the $(K \times K)$-matrix $\mathbf{E}_k$ is all zero except for $[\mathbf{E}_k]_{k,k} = 1$. The $(K \times K)$-matrix $\mathbf{F}_k$ is 1 on its main diagonal and $1/2$ elsewhere, with the k-th column and k-th row being zero:

$$[\mathbf{F}_k]_{i,j} = \begin{cases} 1 & \text{for} \quad i = j \text{ and } j \neq k \\ 1/2 & \text{for} \quad i \neq j \text{ and } i \neq k \text{ and } j \neq k \\ 0 & \text{for} \quad i = k \text{ or } j = k. \end{cases}$$

4.4 Precoding Optimization

The objective for optimization can be various, depending on application, quality of service requirements, channel characteristics, or number of nodes. Common approaches include to maximize either the sum-performance of the network or the performance of a subset of the considered nodes. Here, we consider the fundamental case to guarantee the best performance for the weakest node. This increases the coverage of the sensor network, since nodes with bad channel conditions can gain from nodes with good channel conditions. In other words, the optimization problem we aim to solve can be formulated as follows:

$$\max_{\mathbf{p}_1, \ldots, \mathbf{p}_K} \ \min_{k = 1, \ldots, K} \text{SINR}_k \, , \qquad \text{s.t.} \qquad E_{\text{all}} = 1 \tag{4.4}$$

The given constraint limits the average total transmit energy per symbol-time E_{all}. We find

$$E_{\text{all}} = \frac{1}{f_s} \mathsf{E}\left[\left(\sum_{i=1}^{N} a_i \mathbf{p}_i\right)^T \left(\sum_{i=1}^{N} a_i \mathbf{p}_i\right) + \left(\sum_{i=1}^{N} \bar{a}_i \mathbf{p}_i\right)^T \left(\sum_{i=1}^{N} \bar{a}_i \mathbf{p}_i\right) \right],$$

where the expectation is with respect to the transmit bits a_i and $\bar{a}_i$. Rearranging terms yields

$$E_{\text{all}} = \frac{1}{f_s} \mathsf{E}\left[\sum_{i=1}^{N}\sum_{j=1}^{N} a_i a_j \mathbf{p}_i^T \mathbf{p}_j + \sum_{i=1}^{N}\sum_{j=1}^{N} \bar{a}_i \bar{a}_j \mathbf{p}_i^T \mathbf{p}_j \right]$$

$$= \frac{1}{f_s} \sum_{i=1}^{K}\sum_{j=1}^{K} \mathsf{E}\left[a_i a_j + \bar{a}_i \bar{a}_j\right] \mathbf{p}_i^T \mathbf{p}_j.$$

Note that

$$\mathsf{E}\left[a_i a_j + \bar{a}_i \bar{a}_j\right] = \begin{cases} 2(\frac{1}{2}1^2 + \frac{1}{2}0^2) = 1 & \text{for } i = j \\ 2(\frac{1}{2}1 + \frac{1}{2}0)(\frac{1}{2}1 + \frac{1}{2}0) = \frac{1}{2} & \text{for } i \neq j. \end{cases}$$

Thus, we have

$$E_{\text{all}} = \frac{1}{f_s} \left[\mathbf{p}_1^T, \dots, \mathbf{p}_K^T\right] \begin{bmatrix} \mathbf{I}_N & \frac{1}{2}\mathbf{I}_N & \cdots & \frac{1}{2}\mathbf{I}_N \\ \frac{1}{2}\mathbf{I}_N & \ddots & \ddots & \vdots \\ \vdots & \ddots & \ddots & \frac{1}{2}\mathbf{I}_N \\ \frac{1}{2}\mathbf{I}_N & \cdots & \frac{1}{2}\mathbf{I}_N & \mathbf{I}_N \end{bmatrix} \begin{bmatrix} \mathbf{p}_1 \\ \vdots \\ \mathbf{p}_K \end{bmatrix}.$$

Eventually, the total transmit energy per symbol-time can be written as

$$E_{\text{all}} = \bar{\mathbf{p}}^T (\mathbf{C} \otimes \mathbf{I}_N) \bar{\mathbf{p}}.$$

The (i,j)-th component of the $(K \times K)$-matrix $\mathbf{C}$ is given by

$$[\mathbf{C}]_{i,j} = \begin{cases} T & \text{for } i = j \\ T/2 & \text{for } i \neq j, \end{cases}$$

since $T = \frac{1}{f_s}$.

The objective function SINR_k for node k can be written as a generalized Rayleigh quotient by inclusion of the power constraint. With the substitution

$$\bar{\mathbf{p}} \mapsto \frac{\tilde{\mathbf{p}}}{\sqrt{\tilde{\mathbf{p}}^T(\mathbf{C}\otimes\mathbf{I}_N)\tilde{\mathbf{p}}}},$$

the power constraint is fulfilled for all $\tilde{\mathbf{p}} \in \mathbb{R}^{KN}$:

$$\max_{\bar{\mathbf{p}}\in\mathbb{R}^{KN}} \min_{k=1,\ldots,K} \frac{\bar{\mathbf{p}}^T\left(\mathbf{E}_k\otimes\mathbf{H}_k^T\mathbf{G}\mathbf{H}_k\right)\bar{\mathbf{p}}}{\bar{\mathbf{p}}^T\left(\mathbf{F}_k\otimes\mathbf{H}_k^T\mathbf{G}\mathbf{H}_k\right)\bar{\mathbf{p}}+\sigma_n^2} \quad \text{s.t.} \quad \bar{\mathbf{p}}^T(\mathbf{C}\otimes\mathbf{I}_N)\bar{\mathbf{p}}=1$$

$$= \max_{\bar{\mathbf{p}}\in\mathbb{R}^{KN}} \min_{k=1,\ldots,K} \frac{\frac{\bar{\mathbf{p}}^T}{\sqrt{\bar{\mathbf{p}}^T(\mathbf{C}\otimes\mathbf{I}_N)\bar{\mathbf{p}}}}\left(\mathbf{E}_k\otimes\mathbf{H}_k^T\mathbf{G}\mathbf{H}_k\right)\frac{\bar{\mathbf{p}}}{\sqrt{\bar{\mathbf{p}}^T(\mathbf{C}\otimes\mathbf{I}_N)\bar{\mathbf{p}}}}}{\frac{\bar{\mathbf{p}}^T}{\sqrt{\bar{\mathbf{p}}^T(\mathbf{C}\otimes\mathbf{I}_N)\bar{\mathbf{p}}}}\left(\mathbf{F}_k\otimes\mathbf{H}_k^T\mathbf{G}\mathbf{H}_k\right)\frac{\bar{\mathbf{p}}}{\sqrt{\bar{\mathbf{p}}^T(\mathbf{C}\otimes\mathbf{I}_N)\bar{\mathbf{p}}}}+\sigma_n^2}$$

$$\text{s.t.} \quad \underbrace{\frac{\bar{\mathbf{p}}^T}{\sqrt{\bar{\mathbf{p}}^T(\mathbf{C}\otimes\mathbf{I}_N)\bar{\mathbf{p}}}}(\mathbf{C}\otimes\mathbf{I}_N)\frac{\bar{\mathbf{p}}}{\sqrt{\bar{\mathbf{p}}^T(\mathbf{C}\otimes\mathbf{I}_N)\bar{\mathbf{p}}}}}_{=1}=1$$

$$= \max_{\tilde{\mathbf{p}}\in\mathbb{R}^{KN}} \min_{k=1,\ldots,K} \frac{\tilde{\mathbf{p}}^T\left(\mathbf{E}_k\otimes\mathbf{H}_k^T\mathbf{G}\mathbf{H}_k\right)\tilde{\mathbf{p}}}{\tilde{\mathbf{p}}^T\left(\mathbf{F}_k\otimes\mathbf{H}_k^T\mathbf{G}\mathbf{H}_k+\sigma_n^2\mathbf{C}\otimes\mathbf{I}_N\right)\tilde{\mathbf{p}}}$$

Hence, the optimized precoding vector $\tilde{\mathbf{p}}^*$ can be formulated as

$$\tilde{\mathbf{p}}^* = \arg\max_{\tilde{\mathbf{p}}\in\mathbb{R}^{KN}} \min_{k=1,\ldots,K} \frac{\tilde{\mathbf{p}}^T\mathbf{A}_k\tilde{\mathbf{p}}}{\tilde{\mathbf{p}}^T\mathbf{B}_k\tilde{\mathbf{p}}}. \tag{4.5}$$

The $(KN \times KN)$-matrix $\mathbf{A}_k$ and $\mathbf{B}_k$ in the numerator and in the denominator, respectively, are given by

$$\mathbf{A}_k = \mathbf{E}_k \otimes (\mathbf{H}_k^T\mathbf{G}\mathbf{H}_k)$$
$$\mathbf{B}_k = \mathbf{F}_k \otimes (\mathbf{H}_k^T\mathbf{G}\mathbf{H}_k) + \sigma_n^2\mathbf{C}\otimes\mathbf{I}_N.$$

As shown in [78], the problem (4.5) can be converted to the search of the minimal principal generalized eigenvalue of the convex combination of the matrices $\mathbf{A}_k$ and $\mathbf{B}_k$:

$$\boldsymbol{\mu}^* = \arg\min_{\boldsymbol{\mu}\in[0,1]^K:\|\boldsymbol{\mu}\|_1=1} \lambda_{\max}\left\{\sum_{k=1}^{K}\mu_k\mathbf{A}_k, \sum_{k=1}^{K}\mu_k\mathbf{B}_k\right\}, \tag{4.6}$$

where $\boldsymbol{\mu} = [\mu_1,\ldots,\mu_K]^T$. Note that this problem is quasi-convex and the search over $\boldsymbol{\mu}$ is only of dimension $K-1$. For two nodes the optimization problem results in a line search. This is exemplarily plotted in Fig. 4.2, where $\boldsymbol{\mu} = [\mu_1, 1-\mu_1]^T$. If μ_1 is small, node 2 is

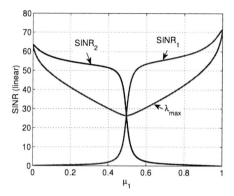

Fig. 4.2: Precoding optimization for two nodes: Example for line search

favored and achieves the better SINR. The reason for this is the larger weight in the combination of the matrices $\mathbf{A}_k$ and $\mathbf{B}_k$. In the same manner, node 1 is preferred when μ_1 is approaching 1. However, in between there is a point where both nodes are served with identical SINR, which solves the max-min optimization problem. A complete proof is given in [78] including a complexity analysis. For a larger number of nodes we used a quasi Newton's method to find μ^*. The optimal transmit pulses are then finally given by

$$\tilde{\mathbf{p}}^* = \begin{bmatrix} \mathbf{p}_1^* \\ \vdots \\ \mathbf{p}_K^* \end{bmatrix} = \mathbf{v}_{\max}\left\{ \sum_{k=1}^{K} \mu_k^* \mathbf{A}_k, \sum_{k=1}^{K} \mu_k^* \mathbf{B}_k \right\}. \tag{4.7}$$

Location-aware channel knowledge: To obtain an SINR expression based on statistical channel knowledge, the expectation of the receiver output (4.2) is taken with respect to the channel. This corresponds to taking the expectation of the numerator and the denominator of the objective function of (4.5) with respect to $\mathbf{H}_k$. In doing so, the structure of the optimization problem is not changed, only the values of matrices in the numerator and denominator are different. They are given by

$$\hat{\mathbf{A}}_k = \mathbf{E}_k \otimes \mathsf{E}_h[\mathbf{H}_k^T \mathbf{G} \mathbf{H}_k]$$
$$\hat{\mathbf{B}}_k = \mathbf{F}_k \otimes \mathsf{E}_h[\mathbf{H}_k^T \mathbf{G} \mathbf{H}_k] + \sigma_n^2 \mathbf{C} \otimes \mathbf{I}_N.$$

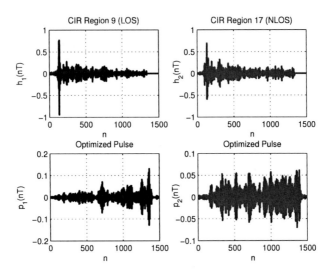

Fig. 4.3: Channel impulse response vectors from region 9 (LOS) and 17 (NLOS) and resulting precoding vectors for two nodes.

As we know from (3.20), we have

$$\mathsf{E}_h[\mathbf{H}_k^T \mathbf{G} \mathbf{H}_k] = \bar{\mathbf{A}}_k = \sum_{n=1}^{N} g_n \mathsf{E}_h\left[[\mathbf{H}_k]_{n,1:N}^T [\mathbf{H}_k]_{n,1:N}\right],$$

where $[\mathbf{H}_k]_{n,1:N}$ denotes the n-th row of $\mathbf{H}_k$ and g_n the n-th element of $\mathbf{g}$. The matrices $\hat{\mathbf{A}}_k$ and $\hat{\mathbf{B}}_k$ depend only on the covariance matrix and mean of the channel impulse response and on the post-detection filter. The optimized transmit pulses based on statistical channel knowledge are then given by (4.6) and (4.7) with $\mathbf{A}_k \mapsto \hat{\mathbf{A}}_k$ and $\mathbf{B}_k \mapsto \hat{\mathbf{B}}_k$.

4.5 Performance Evaluation

To evaluate the performance of the proposed precoding scheme, we simulate the system with measured channel. We use the model as described in Section 4.2 and the precoding as introduced in Section 4.4. As figure of merit, we consider the average bit-error-rate (BER).

The performance evaluation is based on the same setup and measurement campaign as the performance evaluation for single user optimization, see Section 3.5.2. The floorplan

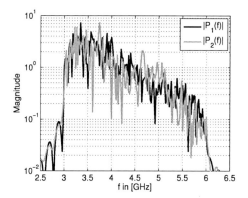

Fig. 4.4: Normalized spectrum of optimized precoding pulses $p_1(t)$ and $p_2(t)$ for region 9 and 17, respectively.

of the measurement environment is plotted in Fig. 3.14. The transmitting central unit is denoted by CH (cluster head), whereas the measurement area is divided into 22 regions of size $27\,\text{cm} \times 56\,\text{cm}$ each. The indoor environment with dense multipath includes line-of-sight (LOS) as well as non-line-of-sight (NLOS) situations. About 600 channel impulse responses are measured per region by moving the receiving antenna on an equidistant horizontal grid. All antenna and hardware effects are assumed to be included in the channel impulse response. The channel impulse responses are aligned to the maximum of their absolute envelope.

We again consider a transmission bandwidth of $B = 3\,\text{GHz}$ at center frequency of $f_c = 4.5\,\text{GHz}$ for the performance evaluation. The symbol timing is $T_{\text{symb}} = 100\,\text{ns}$, which results in a per node data rate of $10\,\text{Mbps}$. To account for stringent complexity requirements of the SNs, the post-detection filter is chosen as first-order low-pass filter with impulse response $g(t)$ given by

$$g(t) = \begin{cases} g_0 \exp\left(-t2\pi f_{\text{cutoff}}\right) & \text{for } t > 0 \\ 0 & \text{else,} \end{cases}$$

with scaling coefficient g_0 and the cutoff frequency set to $f_{\text{cutoff}} = 50\,\text{MHz}$. The bandpass filter at the receiver input is assumed to be perfectly band limiting for the considered transmission band from 3 to 6 GHz. The simulation sampling frequency is chosen to $f_s = 30\,\text{GHz}$, which results in precoding vectors of dimension $N = 1500$.

Fig. 4.3 (top) shows a channel impulse response from a LOS position and one from a

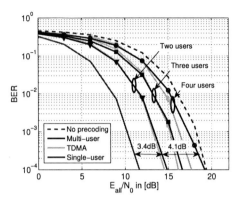

Fig. 4.5: Average bit error rate for precoding with full channel knowledge (Region 9, 17, 20, and 11). Note that E_{all} is the total transmit energy.

NLOS position. Note that we normalize the channel impulse responses to unit energy to suppress the influence of pathloss. Below, the two-user precoding vectors are plotted for two nodes with the corresponding channels. Fig. 4.4 shows the normalized magnitude spectrum of this precoding pulses. The optimization results are for $E_{all}/N_0 = 12$ dB, where E_{all} denotes the average total transmit energy per time-slot and N_0 the noise power spectral density.

First, the performance of multiuser precoding with full channel knowledge is evaluated. We assume static channels for a blocksize of 256 bits, which are randomly drawn from region 9, 17, 20, and 11 with one user in each region. The BER is averaged across channel realizations and users. In Fig. 4.5, the average BER is plotted with precoding optimization for single user and $K = 2, 3$ and 4 (no marker, triangle, square, and circle, respectively). The dashed line plots the average BER without precoding, i.e. a single ideal bandpass pulse of energy E_{all} is used for transmission. The simulations show that the multiuser precoding (blue curves) can effectively orthogonalize the data streams. At high E_{all}/N_0, the performance penalty from single user to two users is only 3.4 dB, and between two and four users it is still only 4.1 dB. Note that the abscissa carries the total energy per time-slot of all data streams. Hence, time division multiplexing (TDMA, cyan) leads to a performance difference of 3.01 dB since E_{all}/N_0 needs to be doubled to serve two users instead of one. This shows that compared to time multiplexing, only about 0.4 dB extra transmit power is necessary, whereas the sum data rate is increased by a factor of two. Moreover, with 1.1 dB additional transmit power, the sum data rate can be quadrupled.

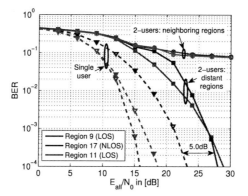

Fig. 4.6: Average bit error rate for precoding regional channel knowledge. Note that E_{all} is the total transmit energy.

For the single user case, the precoding vector is given by $\mathbf{p}^* = \mathbf{v}_{\text{max}}\{\mathbf{A}_1, \mathbf{I}_N\}$. For the multiuser optimization, we use in the SINR expression (4.3) for σ_n^2 the standard deviation of the squared noise term. The simulations show that this leads to the best performance in terms of BER. Thus, we have

$$\sigma_n^2 = 2\left(\mathbf{g}^T(\Sigma_{nn} \odot \Sigma_{nn} - \Sigma_{n'n} \odot \Sigma_{n'n})\mathbf{g}\right)^{1/2},$$

where Σ_{nn} denotes the covariance of the bandpass-filtered noise and $\Sigma_{n'n} = \text{E}[\mathbf{n}_k\mathbf{n}_k'^T]$ the correlation matrix of the noise of the first and the second half of the time-slot, respectively.

Fig. 4.6 shows the averaged BER performance of the precoding with statistical channel knowledge. For each region, the covariance matrix and mean of the channel impulse responses are estimated with the measurement data. The precoding is performed accordingly for different combination of regions. The single user transmission with optimization based on regional channel knowledge is plotted with dashed lines. It can be observed that the performance for the LOS regions 9 (green) and 11 (magenta) is better than for the NLOS region 17 (blue). One explanation for this is the stronger variation of multipath components in the NLOS case, which makes the precoding more difficult. The solid lines show the performance of precoding optimization with two nodes, which are located in different regions. The curves marked with circles correspond to a scenario of two SNs that are located nearby, whereas for the curves marked with squares a setting is chosen with regions further apart. Significant performance difference can be observed. For neighboring regions, the precoding cannot orthogonalize the two data streams and the BER saturates at about 10%. However, for the distant regions 9 and 17 the precoding works well. The SNs can decode their data stream when the transmit power is increased by about 5 dB compared to single user. This shows that multiuser transmission with statistical channel knowledge is possible, if the propagation environment of the SN's regions differs sufficiently, i.e. the SNs are far enough apart.

4.6 Conclusions

In this chapter, we show that precoding for simultaneous transmission to multiple very low-complexity sensor nodes is a very promising means to increase the sum data rate. Optimization algorithms based on full as well as statistical channel knowledge are derived from an SINR expression, where the low complexity receiver structure is specifically taken into account. Performance evaluations based on measurements in a strong multipath environment prove the practicality of the presented scheme.

Chapter 5

Maximum Likelihood Timing Estimation

The following two chapters are the second part of this thesis and contribute to localization in UWB sensor networks. So far, in the first part, we studied the communication and conclude that position information is beneficial for the transceiver adaptation. In the following second part, we focus on the position estimation of SNs. In particular, we investigate timing estimation, which is the key component for range-based localization with UWB impulse radio.

For sensor networks, standard timing estimation algorithms are not applicable due to the stringent requirements on complexity and power consumption of the receiver. Therefore, we present maximum likelihood timing estimation at the output of the generalized energy detection receiver[1]. To account for location-aware channel knowledge, we assume the statistics of the channel to be known at the receiver. To the best of our knowledge, this problem has not been treated in this generalized set up so far. Known approaches for specific post-detection filters rely on a Gaussian approximation of the detector output. We show that the estimation accuracy can be improved by using the exact marginal PDF of the energy detector output. This is verified by experiments with a simple channel model and an extensive measurements campaign. To reduce the complexity, we approximate the energy detector output as multivariate normally or multivariate log-normally distributed random vector. Verification based on the channel model and measured channels favors the log-normal approximation and shows that the correlation is not relevant. Accuracy down to centimeter precision can be achieved.

---

[1]Parts of this chapter have been published in [42].

5.1 Introduction

Timing estimation is a crucial component for range-based UWB localization. Before we present the main contribution of this chapter, we give a short overview on range-based UWB localization to draw the connection to timing estimation. In addition to localization, timing estimation is also necessary for communication. For data transfer, it is generally required that nodes are synchronized. The transmitter and receiver need to agree on a time window for transmission and the receiver needs to find the temporal alignment of the symbols. The basis for synchronization is timing estimation. When the transmitter sends a signal, the receiver has to estimate the beginning of the signal as well as the sampling instance for data detection. The latter is of particular importance for the generalized energy detection receiver, because the wrong sampling time can lead to a strong degradation of the received energy and it can also cause intersymbol interference. In [41], we present a low complexity synchronization scheme for both burst synchronization and symbol synchronization.

For localization, timing estimation enables distance measurements based on ToA or round trip time (RTT) measurements. In the following, we give a brief introduction to localization for low complexity UWB sensor networks. We shortly summarize the ML formulation of the positioning problem based on range measurements. This establishes the link between timing and position estimation. The main contribution of this chapter is the analysis of the timing estimation at the output of the generalized energy detector. In Section 5.2, we introduce the system model and in Section 5.3 we revise the state-of-the-art of conventional ToA estimation. In Section 5.4, we formulate the ML timing estimation problem for the generalized energy detection receiver. The solution is given in Section 5.5 with the derivation of the PDF of the detector output conditioned on the delay. Systematically derived approximations are presented in Section 5.6. The chapter concludes with a performance evaluation presented in Section 5.7 and a summary in Section 5.8.

Related work: Localization with UWB with focus on ToA estimation is described in [17]. In [79], synchronization algorithms for coherent and differential receivers are presented. With focus on low-complexity and energy detection receivers, an algorithm for synchronization has been presented and analyzed in [80]. In [81], maximum likelihood (ML) timing estimators for different IR-UWB receivers are derived, including energy detection receivers. The authors consider a conventional energy detector with rectangular integration window and invoke the Gaussian assumption for the energy detector output.

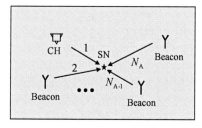 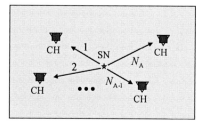

a) Self-localization   b) Remote-localization

Fig. 5.1: Topologies for position estimation in UWB sensor networks

5.1.1 Localization in UWB Sensor Networks

In the context of position estimation, we usually differentiate between agent and anchor nodes. Whereas the anchors are the references, the agents are the nodes whose positions we aim to estimate. The positions of the anchors are known and they are typically stationary and fixed. The anchors are assumed to have a joined clock or to be synchronized with high accuracy. Without loss of generality, we consider the sensor nodes to be the agents. Therefore and to keep the notation simple, the position we want to estimate is $p_{SN} \in \mathbb{R}^3$. The number of available anchors is N_A with their positions given by $p_A^{(i)} \in \mathbb{R}^3$ for $i = 1, \ldots, N_A$. As shown in Fig. 5.1, two different topologies are possible for the position estimation:

- *Self-localization*: The sensor node estimates its own position. The position estimation is based on the downlink channel, i.e. the sensor node is the receiver. Since the anchors are the transmitters, they are not involved in the processing for position estimation. Therefore, besides the CH, simple beacon nodes can be employed as anchors. Since every sensor node determines the position by itself, in literature, self-localization is sometimes also referred to as distributed localization.

- *Remote-localization*: The position estimation takes place at the CH. The localization is based on the uplink channel, i.e. the sensor node is the transmitter and the CH is the receiver. The advantage of this approach is that the signal at the CH achieves higher quality due to loose complexity constraints of the CH. However, this setting requires multiple CHs, since they serve as anchors. The CHs need to disseminate their measurements to determine the position. Alternatively, a single CH with multiple distributed antennas can be used. Since the position estimation is performed always by the CHs, in literature, remote-localization is also referred to as centralized localization.

The choice of the topology has a strong impact on the localization system. On the one hand, it may be preferable to perform remote-localization to keep the complexity of the sensor nodes as low as possible. On the other hand, remote-localization requires a large number of CHs. To achieve a high positioning accuracy, a high number of anchors is necessary. The most beneficial way to increase the positioning accuracy is to increase the number of anchors. Thus, for remote localization the number of CH should be large or they need to have antennas distributed over the whole coverage area of the sensor network. The advantage of self-localization is that simple beacon nodes can be used. They only need to transmit a simple pulse train. The requirements are that their positions are known and that they are synchronized. Since the complexity of the beacon nodes is low, it is reasonable to increase their number to reach a high anchor density in the coverage area of the network. The self-localization requires that the position is estimated from the signals received at the low complexity sensor nodes.

The processing for position estimation is commonly split into several steps. First, some position related parameters are estimated from the receive signals. In our case this is the ToA. Afterwards, the position is estimated from the position related parameters. Generally, this is a suboptimal approach and joint estimation would perform better. The best way would be to find the most probable position that fit to all observed receive signals over all time. However, this approach is too complex in practice since it leads to a problem of very large dimensionality. In this thesis, we consider the following steps to tackle the position estimation problem:

1. Coarse timing estimation (up to the symbol time period)

2. Fine timing estimation (ToA within the symbol period)

3. Position estimation

We differentiate between coarse and fine timing estimation. The coarse estimation aims to find the beginning of a burst, a packet or start and the end of a data transmission. Conventional techniques for coarse timing estimation are based on the RSS or known preambles. In [41], we present and analyze such a data-aided method. The transmitter sends a known sequence and the burst and symbol timing is recovered from the correlation. This enables timing estimation with accuracy up to the symbol period. Other algorithms for coarse timing estimation can be found e.g. in [82] and the references therein.

Fine timing estimation concerns the estimation of the ToA within a symbol period. It is the crucial component to achieve high definition localization. A sensor network with a per node data rate of $10\,\text{Mbps}$ has a symbol period of $100\,\text{ns}$, i.e. the coarse timing estimation

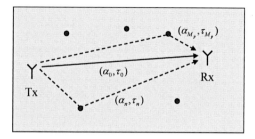

Fig. 5.2: Channel model with LOS path and M_p multipath components

reaches accuracy of up to 30 m. The fine timing estimation is necessary to achieve centimeter accuracy. Therefore, we focus in the following on fine timing estimation of a single transmit pulse $p(t)$ and assume that the coarse estimate is given.

5.1.2 Channel Model for Localization based on Time of Arrival

The channel model gives the connection between the receive signal and the position. Depending on the parameters that are considered for position estimation, the channel model may differ substantially. For the localization based on ToA estimation, we use the channel model as shown in Fig. 5.2. The position information lies in the distance between transmitter and receiver and thus in the delay of the signal. We assume a LOS-path and M_p multipath components from scatterers or reflectors. Each multipath component contributes to the receive signal with a specific delay and amplitude. We assume that the transmitter and receiver are not moving during a signal transmission. Therefore, the delays and amplitudes of the multipath components can be considered to be time-invariant. The receive signal $r'(t)$ that is observed at the receiver, can be written as the superposition of time-shifted transmit signals $s(t)$:

$$r'(t) = \sum_{n=0}^{M_p} \alpha_n s(t - \tau_n) + w(t), \tag{5.1}$$

where α_n and τ_n denote the path gain and path delay corresponding to the n-th path, respectively, and $w(t)$ denotes the noise. Without loss of generally, we assume that α_0 and τ_0

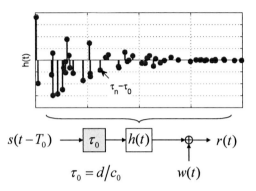

Fig. 5.3: Signal model for localization based on Time-of-Arrival

correspond to the LOS-path. The delay of the LOS-path holds

$$\tau_0 = \frac{d}{c} \tag{5.2}$$

where d is the distance between transmitter and receiver and $c = c_0$ is the speed of light. The delays of the multipath components are always larger than the LOS-delay, i.e.

$$\tau_n \geq \tau_0 \qquad \text{for} \qquad n = 1, \ldots, M_p. \tag{5.3}$$

In this model the position dependence is limited to (5.2) and (5.3). Any other influence of the position on amplitudes and delays is neglected.

For notational convenience, we prefer to exclude the LOS-path delay from the channel impulse response, i.e. we define a channel impulse response $h(t)$ such that it is independent of the distance and aligned to the LOS-path. Thus, the channel impulse response itself has no range information. This is done by writing (5.1) as

$$r'(t) = \delta(t - \tau_0) * \underbrace{\sum_{n=0}^{M_p} \alpha_n \delta(t - \tau_n + \tau_0)}_{:=h(t)} * s(t) + w(t)$$

$$= \delta(t - \tau_0) * h(t) * s(t) + w(t).$$

In most practical cases it is not possible to observe the transmit signal and the receive signal with the same time t. Typically, the transmitter and the receiver have different clocks. Due to complexity and cost requirements of sensor networks, it is not possible to keep the SNs synchronous over a long time. This would require very complex and expensive atomic clocks. Therefore, we assume that the transmitter and receiver are not synchronized and have different clocks. The offset between transmitter-time and receiver-time is denoted by T_0. In addition, the hardware of the transmitter and receiver induces delays from cables and signal processing. The time-shift T_0 includes all processing delays and also time-shifts from the alignment of the channel impulse response. Thus, the receive signal can be modeled by

$$r(t) = \delta(t - \tau_0) * h(t) * s(t - T_0) + w(t)$$
$$= \delta(t - \underbrace{(\tau_0 + T_0)}_{:=\varepsilon}) * h(t) * s(t) + w(t).$$

The receiver cannot directly estimate the distance between transmitter and receiver. Without a joint clock, the receiver sees ε instead of τ_0. Therefore, we define the ToA as

$$\varepsilon := \tau_0 + T_0.$$

5.1.3 Position Estimation

The ToA that corresponds to the anchor k is denoted by $\varepsilon^{(k)}$ with $k = 1, \ldots, N_A$. It holds

$$\varepsilon^{(k)} = \tau_0^{(k)} + T_0 = \frac{1}{c_0} \left\| \mathbf{p}_{\text{SN}} - \mathbf{p}_A^{(k)} \right\| + T_0,$$

where we require T_0 to be independent of the anchor. In literature, the values $\varepsilon^{(k)} \cdot c_0$ are often called pseudo-ranges.

The measured pseudo-ranges are given by $\hat{\varepsilon}^{(k)} \cdot c_0$. With the assumption that the ToA estimation errors are normally distributed and independent, we obtain

$$\hat{\varepsilon}^{(k)} \cdot c_0 \sim \mathcal{N} \left(\left\| \mathbf{p}_{\text{SN}} - \mathbf{p}_A^{(k)} \right\| + c_0 T_0, \sigma_k^2 \right), \qquad k = 1, \ldots, N_A,$$

where σ_k^2 denotes the variance of the k-th pseudo-range.

If transmitter and receiver are not synchronized, the clock offset T_0 is unknown. It must be determined to obtain the distances between agent and anchors and the position of the agent. Typically, the clock offset T_0 is estimated jointly with the position. The ML estimate is formulated as follows:

$$\left\{\hat{\mathbf{p}}_{SN}, \hat{T}_0\right\} = \underset{\mathbf{p}_{SN}\in\mathbb{R}^3, T_0\in\mathbb{R}}{\arg\max} \quad p(\hat{\varepsilon}^{(1)}, \dots, \hat{\varepsilon}^{(N_A)}|\mathbf{p}_{SN}, T_0)$$

$$= \underset{\mathbf{p}_{SN}\in\mathbb{R}^3, T_0\in\mathbb{R}}{\arg\min} \quad \sum_{k=1}^{N_A}\frac{1}{\sigma_k^2}\left(\hat{\varepsilon}^{(k)}\cdot c_0 - \left\|\mathbf{p}_{SN} - \mathbf{p}_A^{(k)}\right\| - c_0 T_0\right)^2.$$

Note that for a given trial position $\mathbf{p}_{SN}$, the minimum must satisfy

$$\frac{\partial}{\partial T_0}\sum_{k=1}^{N_A}\frac{1}{\sigma_k^2}\left(\hat{\varepsilon}^{(k)}\cdot c_0 - \left\|\mathbf{p}_{SN} - \mathbf{p}_A^{(k)}\right\| - c_0 T_0\right)^2 = 0,$$

which results in

$$T_0 = \frac{1}{c_0\bar{\gamma}}\sum_{k=1}^{N_A}\frac{1}{\sigma_k^2}\left(\hat{\varepsilon}^{(k)}\cdot c_0 - \left\|\mathbf{p}_{SN} - \mathbf{p}_A^{(k)}\right\|\right),$$

with $\bar{\gamma} = \sum_{n=1}^{N_A}\frac{1}{\sigma_n^2}$. Thus, we obtain

$$\hat{\mathbf{p}}_{SN} = \underset{\mathbf{p}_{SN}\in\mathbb{R}^3}{\arg\min} \quad \sum_{k=1}^{N_A}\frac{1}{\sigma_k^2}\left(\bar{d}_k - l_k(\mathbf{p}_{SN})\right)^2, \tag{5.4}$$

where

$$\bar{d}_k = \hat{\varepsilon}^{(k)}\cdot c_0 - \frac{1}{\bar{\gamma}}\sum_{n=1}^{N_A}\frac{1}{\sigma_n^2}\hat{\varepsilon}^{(n)}\cdot c_0$$

$$l_k(\mathbf{p}_{SN}) = \left\|\mathbf{p}_{SN} - \mathbf{p}_A^{(k)}\right\| - \frac{1}{\bar{\gamma}}\sum_{n=1}^{N_A}\frac{1}{\sigma_n^2}\left\|\mathbf{p}_{SN} - \mathbf{p}_A^{(n)}\right\|.$$

The solution to (5.4) can be found by iterative algorithms. In [22], an approach is presented based on the Taylor series expansion of the objective function. However, note that the objective function may have several minimums depending on the agent and anchor topology. A necessary condition for a unique solution is that at least four ToA measurements are available, i.e. $N_A \geq 4$.

Under the Gaussian assumption on the ToA estimation error, the position estimation can be separated from ToA estimation.

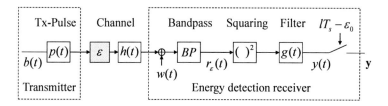

Fig. 5.4: Block diagram of transmitter, channel and receiver

5.2 System Model

For the timing estimation, we consider a UWB impulse radio transmitter, that transmits a sequence of unmodulated pulses $p(t)$, i.e. in Fig. 5.4 we have $b(t) = \sum_n \delta(t - nT_p)$. The pulse repetition period is denoted by T_p. The transmit signal is given by

$$s(t) = \sum_{n=-\infty}^{\infty} p(t - nT_p).$$

Without loss of generality, we omit the data modulation, since we consider non-data-aided timing estimation. The frequency selective fading channel of the multipath environment is characterized by the delay ε and the channel impulse response $h(t)$. For notational convenience, we omit the anchor-index k. Moreover, we assume that the coarse timing estimation is established and the timing offset T_0 is chosen such that ε satisfies

$$0 \leq \varepsilon \leq T_p.$$

Note that this is not a restriction. As it can be seen in (5.4), the position estimate is independent of the absolute value of T_0.

The channel output signal is perturbed by AWGN with power spectral density $N_0/2$. After the receive bandpass we obtain the received signal

$$r_\varepsilon(t) = \sum_{k=-\infty}^{\infty} q(t - kT_p - \varepsilon) + n(t),$$

where $n(t)$ denotes the band-limited noise and $q(t) = \tilde{h}(t) * p(t)$ the receive pulse, which includes channel impulse response, transmit pulse and bandpass filter. We assume that $q(t) = 0$ for $t \notin [0, T_p]$, i.e. no inter pulse interference occurs.

5.3 Conventional Time-of-Arrival Estimators

There exists a large variety of standard ToA estimators for different applications. Most approaches rely on the perfectly sampled receive signal $r_\varepsilon(t)$, which are here referred to as coherent estimators. The implementation requires a high resolution ADC with a very high sampling frequency. Although the coherent estimators are superior in performance, they require huge complexity and power. For sensor networks, coherent estimators can come into consideration only at the CH. Therefore, most standard approaches are limited to the remote-localization. In the following we give a short summary on these conventional approaches for ToA estimation.

Coherent Maximum Likelihood Estimator: The optimal coherent ML estimator is implemented most conveniently in frequency domain. Let us write the receive signal as

$$r_\varepsilon(t) = \sum_{m=0}^{M_p} \alpha_m s(t - \tau_m - T_0) + n(t), \qquad 0 \le t \le T_p.$$

We consider only a single timeslot of length T_p. For longer observation windows, the receive signal can be averaged over the time slots with period T_p, because $r_\varepsilon(t)$ is Gaussian and the correlation can be neglected. After a low-pass filter, sampling with period T and discrete Fourier transform, we obtain

$$R(k) = P(k) \sum_{m=0}^{M_p} \alpha_m e^{j\omega_m k} + N(k), \qquad k = 0, \ldots, N-1,$$

where $\omega_m = -\frac{2\pi(\tau_m + T_0)}{T_p}$ and $N = \frac{T_p}{T}$. The values $P(k)$ and $N(k)$ denote the discrete Fourier transform of the transmit pulse $p(t)$ and the noise $n(t)$, respectively. Since $n(t)$ is a Gaussian process, the noise in frequency domain is normally distributed. The ML estimator becomes

$$
\begin{aligned}
\{\alpha_m^*, & \tau_m^* + T_0\}_{m=1,\ldots,M_p} \\
&= \underset{\alpha_1,\ldots,\alpha_{M_p},\tau_1,\ldots,\tau_{M_p}}{\arg\max} \; p\Big(R(0),\ldots,R(N-1)|\alpha_1,\ldots,\alpha_{M_p},\tau_1,\ldots,\tau_{M_p}\Big) \\
&= \underset{\alpha_1,\ldots,\alpha_{M_p},\tau_1,\ldots,\tau_{M_p}}{\arg\min} \; \sum_{k=0}^{N-1} \left| R(k) - P(k) \sum_{m=0}^{M_p} \alpha_m e^{j\omega_m k} \right|^2 .
\end{aligned}
\tag{5.5}
$$

This nonlinear least squares problem is complicated to solve. However, many close approximations to the solution can be found in literature. In [83], the authors present an efficient algorithm, which is referred to as WRELAX. The ToA is then given by

$$\hat{\varepsilon}_{\text{ML}} = \min_{m=1,\ldots,M_p} \tau_m^* + T_0.$$

Coherent Low-complexity Estimators: Suboptimal estimators can be derived from heuristics or different assumptions on the channel. The following approaches are all based on the matched-filtered receive signal $\tilde{y}(t) = p(-t) * r_\varepsilon(t)$. An example is shown in Fig. 5.5.

- *Max:* The Max-estimator assumes that the LOS-component has the largest path gain and no overlap with the multipath components. Hence, the ToA is given by

$$\hat{\varepsilon}_{\text{max}} = \underset{0 \leq t \leq T_p}{\arg\max} \; |\tilde{y}(t)|$$

- *P-Max:* The P-max-estimator assumes that the LOS-component is within the P-largest path gains and that there is no overlap with the multipath components. The ToA is then given by

$$\hat{\varepsilon}_{\text{p-max}} = \min_{i=1,\ldots,P} \hat{\varepsilon}_{\text{max}}^{(i)},$$

where $\hat{\varepsilon}_{\text{max}}^{(i)}$ denotes the i-th maximum of $|\tilde{y}(t)|$.

- *Simple Thresholding:* Algorithms based on thresholds assume that the magnitude of the noise is bounded and well below the LOS-component. The simple thresholding estimate is given by

$$\hat{\varepsilon}_\eta = \min t \qquad \text{s.t. } |y(t)| > \eta, \; 0 \leq t \leq T_p,$$

where η denotes the threshold. Note that the estimate depends on the amplitude of the LOS-path. For a higher amplitude, the pulse crosses the threshold earlier than for a lower amplitude.

- *Jump Back and Search Forward:* The Jump-Back-and-Search-Forward strategy is a combination of Max-estimator and thresholding. In [47], we propose a simple implementation, which is based on two steps. First, the thresholding is applied in the time window Δ_1 before the maximum:

$$\varepsilon_1 = \min t \qquad \text{s.t. } |y(t)| > \eta_1, \; \varepsilon_{\text{max}} - \Delta_1 \leq t \leq \varepsilon_{\text{max}},$$

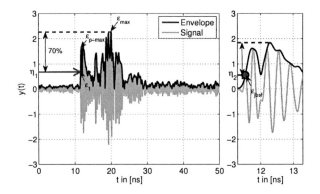

Fig. 5.5: Coherent low-complexity ToA estimator (from [47])

where η_1 is chosen proportional to $|\tilde{y}(\tau_{\max})|$. In the second step, the search is refined using the local maximum in the search window Δ_2:

$$\varepsilon_2 = \underset{\varepsilon_1 \leq t \leq \varepsilon_1 + \Delta_2}{\arg\max} \quad |\tilde{y}(t)|$$

The ToA estimate is then given by

$$\hat{\varepsilon}_{\text{jbsf}} = \min t \qquad \text{s.t.} \ \ |y(t)| > \eta_2, \ \ \varepsilon_2 - \Delta_2 \leq t \leq \varepsilon_2,$$

where η_2 is chosen proportional to $|\tilde{y}(\varepsilon_2)|$.

So far, the state-of-the-art offers powerful algorithms for coherent ToA estimation. This enables precise localization of sensor nodes at the CHs. However, the remote-localization requires many CHs since they serve as anchors. The situation is different for self-localization of the sensor nodes. Whereas the same position estimation algorithm may be used, the requirements on the timing estimation are more stringent. For the coherent timing estimators, the receiver must have access to $r_\varepsilon(t)$, which would require extensive higher complexity for the sensor nodes. Since the sensor nodes are equipped with a generalized energy detection receiver, it is favorable to perform the timing estimation at the output of the receiver. Full rate sampling of the receive signal and complex estimation algorithms are not an option for low complexity nodes. Hence, we propose to estimate the timing at the generalized energy detector output. This way, no further hardware is necessary at the low complexity sensor nodes. In the following, we will derive the ML timing estimator for energy detection receivers.

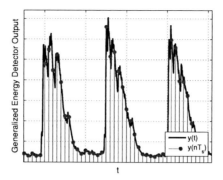

Fig. 5.6: Example for the receive signal at the generalized energy detector output

5.4 Problem Statement

At the low complexity sensor nodes, the received signal $r_\varepsilon(t)$ is detected by the generalized energy detection receiver. It consists of the squaring device and a post-detection filter $g(t)$. The detector output is given by $y(lT_s)$. For communication, we considered only two samples per symbol, i.e. the sampling period was given by $T_s = T_p/2$. For accurate timing estimation it is beneficial to obtain more samples per symbol. This requires slightly higher sampling rates at the receiver output. However, the sampling rate can still be well below the Nyquist rate of the UWB signal, which would be required for coherent timing estimation. An example for the noisy receive signal at the generalized energy detector output is shown in Fig. 5.6, where the solid line shows the continuous-time signal $y(t)$ and the dots the sampled values $y(nT_s)$. The samples are not normally distributed.

In contrast to the conventional timing estimators, we treat the estimation of ε with statistical channel knowledge. We model the channel as a nonstationary Gaussian random process. We assume in the sequel that the receiver does not know $\tilde{h}(t)$, but has full knowledge of the statistics of it, i.e. it knows the mean and the autocorrelation function. Note that $r_\varepsilon(t)$ is a cyclostationary Gaussian random process due to the periodic transmit signal.

The maximum likelihood estimation of ε requires knowledge of the multivariate PDF of the detector output signal $\{y(lT_s)\}$ for any delay ε. There is no closed form solution for this PDF, even though $r_\varepsilon(t)$ is Gaussian. For this reason in many related publications (e.g. [81]) a Gaussian assumption is invoked for $y(t)$.

The delay ε is unknown at the receiver. We treat the estimation of ε under the aforementioned assumptions. As $r_\varepsilon(t)$ is cyclostationary, the estimation range is $[0, T_p]$, i.e. we can only estimate ε mod T_p. The estimation is based on samples taken at time instances $lT_s - \varepsilon_0$ with the unknown sample offset $\varepsilon_0 \in [0, T_s]$ (Fig. 5.4). Thus the estimator actually estimates $\varepsilon_0 + \varepsilon$. As our system is time-invariant, we can incorporate ε_0 into the channel delay ε, i.e. we can assume without loss of generality $\varepsilon_0 = 0$.

Our receiver uses L consecutive T_s-spaced samples of the signal $y(t)$ to generate an estimate $\hat{\varepsilon}$ of the channel delay ε, i.e. the input to the estimator is the vector

$$\mathbf{y} = [y(T_s), y(2T_s), \ldots, y(LT_s)]^T.$$

As known from estimation theory, the maximum likelihood (ML) estimate can be formulated as follows:

$$\hat{\varepsilon} = \arg\max_\varepsilon p_\mathbf{y}(\mathbf{y}|\varepsilon), \tag{5.6}$$

where the PDF of the receive sample vector $\mathbf{y}$ conditioned on the time shift ε is denoted by $p_\mathbf{y}(\mathbf{y}|\varepsilon)$.

In the following, we derive an analytical expression for the marginal PDFs for each energy detector output sample, which are denoted by $p_y(y_n|\varepsilon)$. In order to evaluate $p_\mathbf{y}(\mathbf{y}|\varepsilon)$ for the maximum search, we make the additional assumption that the individual samples $y_l = y(lT_s)$ for $l = 1, \ldots, L$ are statistically independent, which leads to

$$p_\mathbf{y}(\mathbf{y}|\varepsilon) \approx \prod_{l=1}^{L} p_y(y_l|\varepsilon). \tag{5.7}$$

This approximation is necessary, since computation of the multivariate PDF of $\mathbf{y}$ is not feasible using the presented algorithm.

5.5 Marginal PDF of Energy Detector Output for Normally Distributed Channels

For clarity, the following exposition is based on $g(t) = 0$ for $t \notin [0, T_p]$. This is the most practical case and extension to arbitrary $g(t)$ is straightforward. However, note that $g(t) \geq 0$ is required. Due to the cyclostationarity of $y(t)$, we can choose without loss of generality $(lT_s - \varepsilon)$ mod T_p for the calculation of the marginal PDF.

We consider now an equivalent discrete time system model, which is obtained by sampling the continuous signals with sampling period T. All system components are assumed to satisfy a sufficient low-pass characteristic and are approximately time limited that any errors from sampling expansion can be neglected. To account for the squaring operation, the sampling period T must fulfill at least $1/T > 4(B + f_c)$. Then, the sampled generalized energy detector output under the hypothesis that the delay is ε, can be written as

$$y_l = \sum_{n=-N}^{N-1} g(lT_s - \varepsilon - nT)r^2(nT),$$

where $r(t - \varepsilon) = r_\varepsilon(t)$. Thus, the statistics of $r^2(nT)$ are independent of ε. Rewriting this in vector notation yields the quadratic form

$$y_l = \mathbf{r}^T \mathbf{Q}_l^{(\varepsilon)} \mathbf{r}, \tag{5.8}$$

where

$$\mathbf{r} = [r_{-N}, \ldots, r_{N-1}]^T \quad \text{and} \quad \mathbf{Q}_l^{(\varepsilon)} = \text{diag}([g_{-N}, \ldots, g_{N-1}]^T)$$

with $r_n = r(nT)$, $g_n = g(lT_s - \varepsilon - nT)$, and $N = T_p/T$. The diagonal matrix $\mathbf{Q}_l^{(\varepsilon)}$ contains only samples of $g(t)$. Depending on the delay ε and the sample number l, the sampling of $g(t)$ is shifted. In this way, the generalized energy detector output samples can be written compactly, where only $\mathbf{Q}_l^{(\varepsilon)}$ depends on the delay and sampling instance. The vector $\mathbf{r}$ is independent of l and ε and contains the samples of the realization of two receive pulses.

Due to the assumption of Gaussian noise and Gaussian channel taps, the vector $\mathbf{r}$ is jointly Gaussian distributed. The mean $\boldsymbol{\mu}_\mathbf{r}$ of $\mathbf{r}$ is given by

$$\boldsymbol{\mu}_\mathbf{r} = \mathsf{E}(\mathbf{r}) = \mathsf{E}(\tilde{\mathbf{h}}),$$

where $\tilde{\mathbf{h}} = [\tilde{h}(T), \ldots, \tilde{h}(NT), \tilde{h}(T), \ldots, \tilde{h}(NT)]^T$. The noise vector $\mathbf{n}$ is assumed to be zero-mean. The covariance matrix $\boldsymbol{\Sigma}_\mathbf{r}$ of $\mathbf{r}$ is given by

$$\boldsymbol{\Sigma}_\mathbf{r} = \mathsf{E}(\mathbf{r}\mathbf{r}^T) - \boldsymbol{\mu}_\mathbf{r}\boldsymbol{\mu}_\mathbf{r}^T = \mathsf{E}(\mathbf{n}\mathbf{n}^T) + \mathsf{E}(\tilde{\mathbf{h}}\tilde{\mathbf{h}}^T) - \mathsf{E}(\tilde{\mathbf{h}})\mathsf{E}(\tilde{\mathbf{h}})^T.$$

Following the presentation in [25], let us denote λ_i and $\mathbf{u}_i$ as the eigenvalues and the eigenvectors of $\boldsymbol{\Sigma}_\mathbf{r}^{0.5}\mathbf{Q}_l^{(\varepsilon)}\boldsymbol{\Sigma}_\mathbf{r}^{0.5}$. All eigenvalues are nonnegative, due to $g(t) \geq 0$. Furthermore, let

Z_i be independent Gaussian random variables with variance one, and their means given by

$$E\left(Z_i\right) = \gamma_i = \left(\boldsymbol{\Sigma}_{\mathbf{r}}^{-0.5}\mathbf{u}_i\right)^T \boldsymbol{\mu}_{\mathbf{r}}.$$

Then, the quadratic form in (5.8) has the same PDF as the diagonalized version (cf. [84]) given by

$$\tilde{y}_l = \sum_{i=1}^{2N} \lambda_i Z_i^2. \tag{5.9}$$

If all λ_i are equal, y_l is distributed according to a noncentral chi-square distribution. Such a PDF is usually obtained, when the conditional PDF of y_l given a channel realization is sought (e.g. [68]). However, this would only be valid for uniform integration windows, i.e. when $g(t)$ is a rectangular function.

For arbitrary λ_i no closed form expression for $p_y(y_l|\varepsilon)$ is known. Grenander, Pollak, and Slepian presented an efficient and numerically stable approach to calculate $p_y(y_l|\varepsilon)$ for zero mean random variables Z_i ($\gamma_i = 0 \ \forall i$) in [85]. This approach is based on Fourier's inversion of the characteristic function of $\tilde{y}_l$. In [25], this method is extended to the nonzero mean case. The characteristic function of $\tilde{y}_l$ is given by

$$\Psi_{\tilde{y}}(t) = \prod_{i=1}^{2N}(1 - 2t\lambda_i)^{-1/2}\exp\left(\frac{t\gamma_i^2\lambda_i}{1 - 2t\lambda_i}\right).$$

To compute the inverse Fourier transform, we use the identity

$$\frac{\partial\ln\left(\Psi_{\tilde{y}}(t)\right)}{\partial t} = \frac{\partial\Psi_{\tilde{y}}(t)}{\partial t}\frac{1}{\Psi_{\tilde{y}}(t)}$$

and obtain

$$\frac{\partial\Psi_{\tilde{y}}(t)}{\partial t}\frac{1}{\Psi_{\tilde{y}}(t)} = \sqrt{-1}\sum_{i}^{2N}\frac{\lambda_i}{1 - 2t\lambda_i} + \frac{\gamma_i^2\lambda_i}{(1 - 2t\lambda_i)^2}. \tag{5.10}$$

Since $p_{\tilde{y}}(\tilde{y}_l|\varepsilon) = p_y(y_l|\varepsilon)$, the inverse Fourier transform of (5.10) yields

$$y_l \cdot p_y(y_l|\varepsilon) = \int_0^{y_l} p_y(y_l - \tau|\varepsilon)a(\tau)\mathrm{d}\tau, \tag{5.11}$$

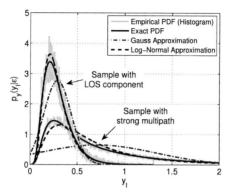

Fig. 5.7: Exact PDF of generalized energy detector output sample compared to histogram and closed form approximations

where

$$a(\tau) = \frac{1}{2} \left(\sum_{i=1}^{2N} \exp\left(-\frac{\tau}{2\lambda_i}\right) \left(1 + \frac{\gamma_i^2 \tau}{2\lambda_i}\right) \right).$$

Eq. (5.11) can be solved numerically by trapezoidal integration, see [39,85]. Thus, the exact PDF of the energy detector output samples for arbitrary Gaussian processes at the input and arbitrary non-negative integration filters can be computed. This enables the evaluation of the maximum likelihood rule given by (5.6).

Fig. 5.7 depicts the numerically obtained PDF for two different scenarios. First, the PDF is plotted for a sample with significant mean and less variance. This corresponds, e.g., to a sample with strong line-of-sight (LOS) component. Furthermore, the PDF of a sample with less mean and larger variance is plotted. This is related to a sample containing strong multipath components. In addition to the exact PDF, an empirical PDF is plotted. This is obtained by computing the histogram of energy detector processed channel and noise realizations drawn from the normal distribution. Furthermore, the normal and log-normal distributions with same mean and variance are plotted.

5.6 Low-complexity Approximations of PDF

Even though the numeric evaluation of $p_y(y_l|\varepsilon)$ might be fast using the algorithm presented in [25], for practical system implementation it may be too complex. Therefore, we present approximations for $p_\mathbf{y}(\mathbf{y}|\varepsilon)$ that allow to solve the ML-estimation with lower complexity. First, we follow the approach to approximate the distribution of $\mathbf{y}$ by $\mathcal{N}(\boldsymbol{\mu}_\mathbf{y}, \boldsymbol{\Sigma}_\mathbf{y})$. The timing estimation corresponds then to the well-known timing estimators as used for linear receivers [86]. Second, we approximate the generalized energy detector output as log-normally distributed. Both approximations are shown in Fig. 5.7 in comparison to the exact PDF. It can be seen that the log-normal distribution approximates the PDF better than the Gauss approximation.

5.6.1 Gaussian Approximation

With the approximation $p_\mathbf{y}(\mathbf{y}|\varepsilon) \approx \mathcal{N}(\boldsymbol{\mu}_{\mathbf{y},\varepsilon}, \boldsymbol{\Sigma}_{\mathbf{y},\varepsilon})$ the ML-estimator (5.6) simplifies to

$$\hat{\varepsilon}_{\text{Gauss}} = \arg\min_\varepsilon (\mathbf{y} - \boldsymbol{\mu}_{\mathbf{y},\varepsilon})^T \boldsymbol{\Sigma}_{\mathbf{y},\varepsilon}^{-1} (\mathbf{y} - \boldsymbol{\mu}_{\mathbf{y},\varepsilon}). \tag{5.12}$$

The mean and covariance matrix are given by

$$\boldsymbol{\mu}_{\mathbf{y},\varepsilon} = \mathbf{G}_\varepsilon \boldsymbol{\mu}_{\mathbf{r}^2} \quad \text{and} \quad \boldsymbol{\Sigma}_{\mathbf{y},\varepsilon} = \mathbf{G}_\varepsilon \boldsymbol{\Sigma}_{\mathbf{r}^2} \mathbf{G}_\varepsilon^T.$$

where $\mathbf{G}_\varepsilon \in \mathbb{R}^{L \times N}$ with

$$[\mathbf{G}_\varepsilon]_{l,n} = g(\text{mod}(lT_s - \varepsilon - nT, T_p))).$$

Further, $\boldsymbol{\mu}_{\mathbf{r}^2} \in \mathbb{R}^N$ and $\boldsymbol{\Sigma}_{\mathbf{r}^2} \in \mathbb{R}^{N \times N}$ are mean vector and covariance matrix after the squaring device, respectively, with

$$[\boldsymbol{\mu}_{\mathbf{r}^2}]_n = \mathsf{E}\left[r^2(nT)\right] = \sigma_{nn} + \mu_n^2$$
$$s_{i,j} = \mathsf{E}\left[r^2(iT)r^2(jT)\right] - \mathsf{E}\left[r^2(iT)\right]\mathsf{E}\left[r^2(jT)\right]$$
$$= \sigma_{ii}\sigma_{jj} + 2\sigma_{ij}^2 + 4\mu_i\mu_j\sigma_{ij} + \mu_i^2\sigma_{jj} + \mu_j^2\sigma_{ii} + \mu_i^2\mu_j^2,$$

where μ_i is the i-th entry of $\boldsymbol{\mu}_\mathbf{r}$ and σ_{ij} is the (i,j)-th entry of $\boldsymbol{\Sigma}_\mathbf{r}$.

The next step towards lower complexity is to neglect correlation in the elements of $\mathbf{y}$, i.e. approximating all off-diagonal entries of the covariance matrix with zero. The ML-rule is then

$$
\begin{aligned}
\hat{\varepsilon}_{\text{Gauss,diag}} &= \arg\min_{\varepsilon} \sum_{l=1}^{L} \frac{\left([\mathbf{y}]_l - \left[\boldsymbol{\mu}_{\mathbf{y},\varepsilon}\right]_l\right)^2}{[\boldsymbol{\Sigma}_{\mathbf{y},\varepsilon}]_{l,l}} \\
&= \arg\max_{\varepsilon} \sum_{l=1}^{L} \frac{[\mathbf{y}]_l \left[\boldsymbol{\mu}_{\mathbf{y},\varepsilon}\right]_l - \left[\boldsymbol{\mu}_{\mathbf{y},\varepsilon}\right]_l^2}{[\boldsymbol{\Sigma}_{\mathbf{y},\varepsilon}]_{l,l}}
\end{aligned}
\tag{5.13}
$$

Moreover, if the variance of all output samples is approximated to be constant, i.e. $\boldsymbol{\Sigma}_{\mathbf{y},\varepsilon} \approx \sigma_y^2 \mathbf{I}$, the estimator simplifies to

$$
\hat{\varepsilon}_{\text{Gauss},\Sigma=\mathbf{I}} = \arg\max_{\varepsilon} \mathbf{y}^T \boldsymbol{\mu}_{\mathbf{y},\varepsilon}.
\tag{5.14}
$$

This corresponds to correlation of the received samples with the expected value and is similar to the estimator proposed in [81].

The estimator can be further simplified by the approximation

$$
\boldsymbol{\mu}_{\mathbf{y},\varepsilon} \approx [0, \ldots, 0, \mu_{k_0}, 0, \ldots, 0]^T
$$

which yields

$$
\hat{\varepsilon}_{\max(\mathbf{y})}/T = \arg\max_{k} \mathbf{y}^T [k - k_0].
\tag{5.15}
$$

In this case, the largest value of the receive vector is taken which might be the least complex timing estimation. This approach has been investigated in detail in [80].

5.6.2 Log-Normal Approximation

The log-normal distribution is the PDF of a random variable whose logarithm is normally distributed. The multivariate log-normal approximation of the PDF $p_{\mathbf{y}}(\mathbf{y}|\varepsilon)$ is given by

$$
p_{\mathbf{y}}(\mathbf{y}|\varepsilon) \approx \frac{\exp\left(-\frac{1}{2}\left(\ln(\mathbf{y}) - \boldsymbol{\zeta}_{\varepsilon}\right)^T \mathbf{V}_{\varepsilon}^{-1} \left(\ln(\mathbf{y}) - \boldsymbol{\zeta}_{\varepsilon}\right)\right)}{(2\pi)^{L/2} |\mathbf{V}_{\varepsilon}|^{1/2} \prod_{l=1}^{L} y(lT_s + \varepsilon)},
$$

for $y_l > 0$ and 0 otherwise. This approximation corresponds to computing element-wise the logarithm of the generalized energy detector output and assuming this to be $\mathcal{N}(\boldsymbol{\zeta}_{\varepsilon}, \mathbf{V}_{\varepsilon})$ distributed [87]. The estimation rule based on the multivariate log-normal approximation yields

$$\hat{\varepsilon}_{\text{MV}-\text{LN}} = \arg\min_{\varepsilon} \, (\ln(\mathbf{y}) - \boldsymbol{\zeta}_{\varepsilon})^T \mathbf{V}_{\varepsilon}^{-1} (\ln(\mathbf{y}) - \boldsymbol{\zeta}_{\varepsilon}). \tag{5.16}$$

The parameters $\boldsymbol{\zeta}_{\varepsilon} \in \mathbb{R}^L$ and $\mathbf{V}_{\varepsilon} \in \mathbb{R}^{L \times L}$ are given by

$$[\boldsymbol{\zeta}_{\varepsilon}]_l = \ln\left(\left[\boldsymbol{\mu}_{\mathbf{y},\varepsilon}\right]_l\right) - \frac{1}{2}\ln\left(1 + \frac{[\boldsymbol{\Sigma}_{\mathbf{y},\varepsilon}]_{l,l}}{\left[\boldsymbol{\mu}_{\mathbf{y},\varepsilon}\right]_l^2}\right)$$

$$[\mathbf{V}_{\varepsilon}]_{i,j} = \ln\left(1 + \frac{[\boldsymbol{\Sigma}_{\mathbf{y},\varepsilon}]_{i,j}}{\left[\boldsymbol{\mu}_{\mathbf{y},\varepsilon}\right]_i \left[\boldsymbol{\mu}_{\mathbf{y},\varepsilon}\right]_j}\right).$$

Neglecting the correlation of the energy detector output samples, the distribution can be approximated by univariate log-normal. In this case, the estimation rule simplifies to

$$\hat{\varepsilon}_{\text{LN}} = \arg\max_{\varepsilon} \sum_{l=1}^{L} \frac{-(\ln([\mathbf{y}]_l) - [\boldsymbol{\zeta}_{\varepsilon}]_l)^2}{[\mathbf{V}_{\varepsilon}]_{l,l}}. \tag{5.17}$$

An advantage of the log-normal approach is the very efficient approximation of the logarithm in a VLSI implementation.

5.7 Performance Results

This section presents the performance evaluation of the timing estimators. First, the performance is evaluated based on channel realizations drawn randomly from normal distribution with exponential power delay profile (PDP). Second, the performance is evaluated based on measured channel impulse responses (CIRs). As the figure of merit, we consider the root mean square (RMS) of the estimation error, which is obtained by Monte Carlo simulation. To find the ML-estimates, we perform an exhaustive search over ε. This is done by discretizing the time shift with step size Δ, yielding $\varepsilon_k = \Delta \cdot k$. The likelihood $p_{\mathbf{y}}(\mathbf{y}|\varepsilon_k = \Delta \cdot k)$ is determined for all $k = 1, \ldots, K$ with $K = T_p/\Delta$. Then, $\hat{\varepsilon}$ is chosen corresponding to the largest likelihood value. Simulation results show that all estimators are unbiased.

System parameters are chosen as follows. The bandwidth is $B = 3$ GHz, using a center frequency of $f_c = 4.5$ GHz. The UWB pulses are band-limited and flat in frequency in the desired bandwidth from 3 GHz to 6 GHz. The pulse repetition period is chosen to be $T_{\text{symb}} = 25$ ns. The sampling frequency as well as the rate of the step size to discretize ε is

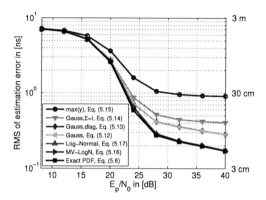

Fig. 5.8: Performance of timing estimators for normally distributed CIRs with exponential PDP

$1/T = 1/\Delta = 30$ GHz. For the estimation, the samples of one pulse repetition period T_{symb} are taken into account, i.e. the number of collected samples is $L = T_{\text{symb}}/T_s$.

According to the model, the performance is evaluated for channel realizations drawn randomly from a distribution $\mathcal{N}(0, \Sigma_{\tilde{h}})$. The covariance matrix $\Sigma_{\tilde{h}}$ is chosen correspondingly to an exponential PDP with RMS delay spread of 3 ns and band limitation to the spectrum from 3 to 6 GHz. The post-detection filter $g(t)$ is chosen as a rectangular function with 1 ns duration and the sampling frequency at the output of the energy detector is set to $1/T_s = 1$ GHz. Fig. 5.8 shows the performance of the timing estimators (5.6) and (5.12)-(5.17). The RMS of the estimation error is plotted versus the signal-to-noise ratio E_p/N_0, with E_p denoting the energy per pulse and $\frac{N_0}{2}$ the noise power spectral density. The error is given in nanoseconds as well as meters, when assuming a propagation speed of $c_0 = 3 \cdot 10^8$ m/s. For all estimators, the error of the timing estimate saturates for increasing E_p/N_0 due to the remaining uncertainty in the channel. As expected, the performance increases with increasing estimator complexity. However, with this choice of integration window and sampling at the energy detector output, the correlation between the samples is low. Therefore, the estimators neglecting the correlations perform very similar to the ones considering the multivariate PDF. The log-normal approximation reaches nearly the performance of the estimator using the exact PDF. This shows that the energy detector output samples are well modeled by log-normal distribution in this case.

Second, the performance is evaluated for measured CIRs, cf. Fig. 5.9. Two different scenarios are chosen, one typical LOS situation and one non-line-of-sight (NLOS) situation

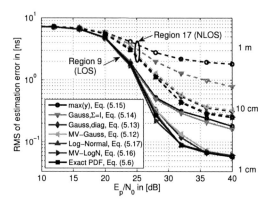

Fig. 5.9: Performance of Timing Estimators for LOS and NLOS Channel

in an indoor environment with rich multipath propagation. For each scenario, the parameters $\mu_{\tilde{h}}$ and $\Sigma_{\tilde{h}}$ are estimated from a set of 620 measured CIRs, that are obtained by moving the receiver over an area of 27×56 cm. The measurement setup and postprocessing is described in detail in [74]. For the timing estimation the channel is chosen randomly out of the set of measured CIRs. To omit the influence of path loss the energy of each CIR is normalized. To account for the low complexity requirements of the receiver, we choose now a first-order low-pass filter after the squaring device, i.e.

$$g(t) = e^{-t2\pi f_{\text{cutoff}}} \text{ for } t > 0 \text{ and } g(t) = 0 \text{ else}, \tag{5.18}$$

with cutoff frequency $f_{\text{cutoff}} = 300$ MHz.

Some interesting observations can be made from the simulation results. For the LOS channel characteristics all timing estimators perform almost similar well. Only marginal improvement can be seen for more accurate modeling of the PDF in this case. In contrast, the NLOS scenario shows larger difference in performance for the different estimators. The estimate based on (5.15) shows as expected the worst performance and does not give anymore a reliable timing estimate. Also the estimator based on Gauss approximation with equal variance of all energy detector output samples shows very poor performance. Large improvement can be seen when the variance of the different energy detector output samples is taken into account, cf. (5.13). However, this performs almost equally well as the estimator that considers also the off-diagonal elements of the covariance matrix, cf. (5.12). Again, the estimators based on the log-normal approximation shows great performance.

5.8 Summary

Maximum likelihood estimators for timing estimation of the receive pulse for UWB impulse radio have been investigated. Based on the assumption of normally distributed channels, the PDF of the energy detector output is derived. Moreover, closed form approximation based on the normal and log-normal distribution are given. Performance evaluation shows that in LOS situations even low complexity timing estimation performs well. For NLOS scenarios the log-normal approximation shows promising results. Moreover, it is easy to implement by taking the logarithm of the energy detector output. Then, the values have to be scaled according to the variance and correlated with the expected value.

Chapter 6

Spectral Timing Estimation

In this chapter[1], we present a receiver structure for ToA estimation, which can be implemented with very low complexity and low power consumption. Standard approaches for ToA estimation of UWB pulses require very high speed A/D conversion, which makes them impractical for implementation. Although ToA estimation based on phase estimation of narrowband signals circumvents that problem, the estimation performance suffers strongly from multipath propagation. We propose to combine the advantages of both techniques by performing spectral estimation at the output of a UWB energy detection receiver. This system benefits from resolvability of multipath components due to a large signaling bandwidth, but requires only low sampling rates. The impact of bandwidth on the ToA estimation accuracy is derived analytically and a closed form approximation of the estimation variance for a multipath channel is given. Moreover, we present a performance evaluation based on measured channels and show that accuracy up to 20 cm can be reached in strong multipath environments.

6.1 Introduction

The state-of-the-art of UWB localization systems requires receivers that are too power hungry and complex for extensive mass-market deployment. In this chapter, we aim for a low complexity, low power and low cost positioning system that still offers reasonable localization accuracy, as it is necessary for location-aware wireless sensor networks. The typical application environment is indoor and is characterized by rather short distances of up to about 10 m.

---

[1]Parts of this chapter have been published in [43].

91

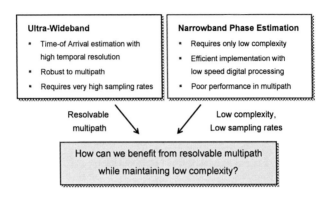

Fig. 6.1: Ultra-Wideband versus Narrowband position estimation

In terms of low complexity and power consumption, narrowband positioning systems, which are based on phase measurements, are the method of choice (e.g. [88], [89]). In an indoor environment, the multipath propagation is a major source of impairment however and leads to substantial positioning errors [19]. UWB positioning systems can resolve the multipath and are as such much more robust [17]. They require huge sampling rates however and are thus not directly suitable for our application regime. The key idea of this chapter is to combine aspects of narrowband and UWB positioning such that the advantages of both approaches are captured, see Fig. 6.1.

Before we summarize the contribution of the chapter, a short overview of related work in UWB and narrowband ToA estimation is given. For a comprehensive review of indoor position location techniques, the reader is referred to [19] and [17]. Reference [82] provides an overview of timing and ToA estimation for UWB receivers. In [20], the generalized maximum-likelihood estimator for the direct path in presence of multipath components is presented and analyzed. Reference [90] presents threshold-based ToA estimators. Even though these estimators are superior in terms of performance, they are impractical for implementation, because they require Nyquist sampling of the received signal. When using the whole UWB bandwidth according to FCC regulation this would require an ADC of at least 15 GHz sampling rate. As shown in the previous chapter, the sampling rate can be decreased by using noncoherent receivers. However, this still requires digital processing in the gigahertz range. In contrast to this, narrowband positioning systems based on phase measurements including carrier phase-based differential GPS (DGPS) can be implemented requiring ADC and digital processing only with about or less than 2 MHz sampling rate [89].

However, when applied in a multipath indoor environment the phase of the narrowband signal becomes random. The single multipath contributions are not resolvable which leads to substantial positioning errors.

In this chapter, we propose and analyze a hybrid system, which combines elements of UWB and narrowband positioning. We assume a UWB impulse radio (IR) source, which generates either a dedicated training sequence or a data signal. Thus, our system can be implemented either as stand-alone positioning system or as low complexity add-on to the communication in the UWB sensor network. For complexity reasons, we consider a non-data-aided ToA estimator without any channel state information. Specifically, the received signal is processed by the standard generalized energy detector front-end. In the add-on scenario, this front-end is present anyway for data detection. The ToA specific part of the receiver then consists of a narrowband quadrature demodulator, low-pass and low rate ADC. Our ToA estimator is based on the observation that the energy detector output signal has a periodic component and thus discrete spectral lines at multiples of the symbol frequency. The ToA estimate is deduced from the complex amplitude of the spectral line at symbol frequency. In this respect, the estimator is closely related to the classical square timing recovery [91].

Besides the basic idea to combine UWB and narrowband positioning, the main contribution of the chapter is an analysis of this estimator under multipath propagation. Simulation results of the estimator performance for simulated and measured channel impulse responses confirm the theory and highlight the improved multipath robustness.

The remainder of the chapter is structured as follows. In Section 6.2, the considered system model is described. In Section 6.3, the input/output relation of the proposed receiver structure is derived. The impact of multipath propagation is analyzed in Section 6.4. Simulation results on the estimator performance are given in Section 6.5.

6.2 System Model

We consider a radio transmitter that transmits a sequence of modulated pulses $p(t)$, i.e. in Fig. 6.2 we have

$$b(t) = \sum_k \tilde{a}_k \delta(t - kT_p).$$

The pulse repetition period is denoted by T_p. For the add-on scenario the transmit symbols $\tilde{a}_k$ carry the data stream. In the case of binary OOK, the mapping of the input bit to the

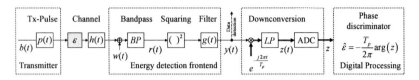

Fig. 6.2: Block diagram of transmitter, channel and receiver

transmit symbol is given by "0" $\mapsto 0$ and "1" $\mapsto (\pm 1)$ with $T_p = T_{\text{symb}}$. Note that the polarity of the "1"-symbol is chosen randomly to avoid discrete spectral lines in the transmit signal spectrum. Otherwise, the FCC rules for UWB could not be exploited to the full extent. For BPPM in turn each input bit corresponds to two output symbols, i.e. we use the mapping "0" $\mapsto (\pm 1, 0)$ and "1" $\mapsto (0, \pm 1)$ with $T_p = T_{\text{symb}}/2$. We again use random pulse amplitudes (± 1), which are independent and identically distributed (i.i.d.) such that $\mathrm{E}[\tilde{a}_k] = 0$ and spectral lines are avoided. For the stand-alone scenario the symbols $\tilde{a}_k$ do not carry a data stream and are i.i.d. for the same reason. Note that in all cases the ToA estimator does not require knowledge of the transmit symbols $\tilde{a}_k$.

The transmit signal is given by

$$s(t) = \sum_{k=-\infty}^{\infty} \tilde{a}_k p(t - kT_p),$$

which is transmitted over the propagation channel with delay ε and impulse response $h(t)$. At the receiver, the signal is perturbed by additive white Gaussian noise $w(t)$. First, the signal is processed by the generalized energy detection receiver. It consists of the bandpass filter, a squaring device and a post detection filter with impulse response $g(t)$. The output $y(t)$ is then downconverted by mixing with frequency $1/T_p$. Note that all this processing can be done in analog with low complexity. In the digital part, the complex value of the DC ($f = 0$) component of the low-pass output is estimated. Thus, the sampling rate of the ADC is determined by the desired number of ToA estimates per second. It is independent of other system parameters, in particular the pulse rate and the system bandwidth.

The ToA estimate is obtained by computing the phase according to

$$\hat{\varepsilon} = -\frac{T_p}{2\pi} \arg(z), \tag{6.1}$$

where $\arg(\cdot)$ gives the argument of the complex number. For the analysis of the ToA estimation, we impose the following assumptions:

- The local oscillator can produce the exact frequency $1/T_p$ and does not suffer from any phase noise or frequency offset.

- There is no interference from other nodes or other wireless systems.

- The base pulse $p(t)$ and impulse responses of $g(t)$, the bandpass and low-pass are even, i.e. their spectra are real.

Once the ToA is estimated from different anchors, the nodes can determine their position from the pseudo-ranges by joint position and clock-offset estimation, see Section 5.1.3.

6.3 Input Output Relation

First, we derive the input/output relation of the considered receiver structure. The signal at the output of the bandpass filter is given by

$$r(t) = \sum_{k=-\infty}^{\infty} \tilde{a}_k q(t - kT_p) + n(t),$$

where $n(t)$ denotes the band-limited noise and $q(t)$ the receive base pulse, which comprises transmit pulse shape, delay ε, channel impulse response, and bandpass filter.

After squaring and post-detection filtering by the filter with impulse response $g(t)$, we obtain the signal $y(t)$. An example of such a signal is shown in Fig. 6.3 together with its magnitude spectrum $Y(f)$. The signal $y(t)$ consists of a periodic component $y_P(t)$ and a non-periodic component $y_{NP}(t)$, i.e. $y(t) = y_P(t) + y_{NP}(t)$. The periodic component $y_P(t)$ yields the discrete spectral lines, which we are interested in. It is given by

$$y_P(t) \hat{=} \mathsf{E}[y(t)] = \mathsf{E}\left[g(t) * \left(\sum_{k=-\infty}^{\infty} \tilde{a}_k q(t - kT_p) + n(t) \right)^2 \right]$$

$$= g(t) * \sum_{k=-\infty}^{\infty} \mathsf{E}[\tilde{a}_k^2] \cdot q^2(t - kT_p) + \sigma_g^2, \tag{6.2}$$

where the constant mean of the squared noise term is denoted by σ_g^2. Eq. (6.2) holds if at least one of the following conditions applies:

i) $\mathsf{E}[\tilde{a}_k \tilde{a}_l] = \delta[k - l],$ \hfill (6.3)

ii) $q(t) \cdot q(t - kT_p) = 0, \forall k \neq 0.$ \hfill (6.4)

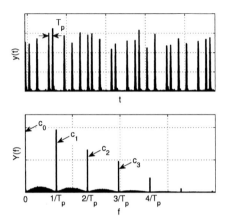

Fig. 6.3: Example of energy detector output signal $y(t)$ and magnitude spectrum $Y(f)$.

The first condition holds for both the add-on and the stand-alone scenario, if the symbol mapping is done as described in Section 6.2. Eq. (6.4) implies non-overlapping receive impulses, i.e. no inter-pulse interference is considered. This condition is frequently met in low to medium data rate systems (pulse period much larger than pulse duration) as they are typical for wireless sensor networks.

The non-periodic component $y_{NP}(t)$ is due to the additive noise and the random transmit symbols $\tilde{a}_k$. It causes an increase of the estimation error variance. Note that this effect can be controlled by the low-pass filter bandwidth. In contrast, the estimation error due to the multipath propagation is independent of the filter bandwidth. For this reason we focus on the multipath scenario in the sequel and assume $y_{NP}(t) = 0$ and $\sigma_g^2 = 0$. As our positioning system is based on the periodic component, we will analyze the impact of multipath propagation on $y_P(t)$ and on the estimator performance.

In frequency domain, the energy detector output (6.2) can then be written equivalently

$$Y(f) = G(f)\frac{1}{T_p} \sum_{k=-\infty}^{\infty} c_k\delta\left(f - \frac{k}{T_p}\right),$$

where $G(f)$ denotes the transfer function of the post-detection filter and c_k the Fourier series coefficients with value

$$c_k = \int\limits_{-\infty}^{\infty} Q(\nu) Q\left(\tfrac{k}{T_p} - \nu\right) d\nu.$$

The spectrum of the equivalent receive base pulse $q(t)$ is denoted by $Q(\nu)$.

Downconversion and low-pass filtering cuts out the Fourier coefficient c_1. The spectrum is then given by

$$Z(f) = G\left(\tfrac{1}{T_p}\right) \frac{1}{T_p} \delta(f) \int\limits_{-\infty}^{\infty} Q(\nu) Q\left(\tfrac{1}{T_p} - \nu\right) d\nu$$

or in time domain

$$z(t) = G\left(\tfrac{1}{T_p}\right) \frac{1}{T_p} \int\limits_{-\infty}^{\infty} Q(\nu) Q\left(\tfrac{1}{T_p} - \nu\right) d\nu. \tag{6.5}$$

6.4 Multipath Environment

This section presents a thorough analysis of the ToA estimator according to (6.1) in an environment with multipath propagation. In particular, the impact of the bandwidth of the base pulse on the multipath distortion is investigated.

First, we consider the situation that the system is applied in a direct LOS situation between transmitter and receiver antenna, i.e. without any multipath components. In this case, the channel induces only the desired delay of the transmit signal due to the time-of-flight according to the distance. Hence, the impulse response of the channel is given by

$$h_{\text{LOS}}(t) = \alpha_0 \cdot \delta(t),$$

where α_0 denotes the attenuation of the radio signal. Together with the delay ε, the receive pulse becomes

$$q_{\text{LOS}}(t) = \alpha_0 \tilde{p}(t - \varepsilon),$$

where $\tilde{p}(t)$ denotes the bandpass filtered transmit pulse $p(t)$. In frequency domain, we have

$$Q_{\text{LOS}}(f) = \alpha_0 e^{-j2\pi f \varepsilon} \tilde{P}(f).$$

Substitution into (6.5) leads to

$$z_{\mathrm{LOS}}(t) = G\left(\tfrac{1}{T_p}\right)\frac{1}{T_p}\int\limits_{-\infty}^{\infty}\alpha_0 e^{-j2\pi\nu\varepsilon}\tilde{P}(\nu)\cdot\alpha_0 e^{-j2\pi\frac{\varepsilon}{T_p}+j2\pi\nu\varepsilon}\tilde{P}\left(\tfrac{1}{T_p}-\nu\right)d\nu$$

$$= G(\tfrac{1}{T_p})\frac{\alpha_0^2}{T_p}\cdot e^{-j2\pi\frac{\varepsilon}{T_p}}\int\limits_{-\infty}^{\infty}\tilde{P}(\nu)\tilde{P}\left(\tfrac{1}{T_p}-\nu\right)d\nu.$$

By assumption, the spectrum $\tilde{P}(f)$ of the bandpass filtered transmit pulse is real. The phase gives then the desired result:

$$\arg z_{\mathrm{LOS}}(t) = -2\pi\frac{\varepsilon}{T_p}+2\pi\kappa, \quad \kappa\in\mathbb{Z}.$$

This way, the timing estimator according to (6.1) is motivated. The ambiguity with κ translates to the ranging ambiguity of $T_p\cdot c$.

6.4.1 Channel Model

We apply now a channel modeling approach with one direct LOS path and M_p paths from scatterers or reflectors. Hence, the channel impulse response is given by

$$h(t) = \sum_{n=0}^{M_p}\alpha_n\delta(t-\gamma_n), \tag{6.6}$$

where α_n denote the path gain corresponding to the n-th path. The path delays[2] γ_n are aligned to the LOS-path, which has the index 0, i.e. $\gamma_0 = 0$ and $\gamma_n \geq 0$ for $n = 1,\ldots,M_p$. The receive pulse is then given by

$$q(t) = \sum_{n=0}^{M_p}\alpha_n\tilde{p}(t-\varepsilon-\gamma_n).$$

---

[2]Note that (6.6) is in line with the definition in Section 5.1.2 for $\gamma_n := \tau_n - \tau_0$ and $\varepsilon = \tau_0 + T_0$. To keep the notation as short as possible, we prefer to use γ_n instead of $\tau_n - \tau_0$.

6.4.2 Transmit Pulse

For the ease of analysis, we choose the transmit base pulse as an ideal low-pass signal. The spectrum of the base pulse is thus given by

$$\tilde{P}(f) = \begin{cases} \frac{1}{\sqrt{B}} & \text{for } -\frac{B}{2} < f < \frac{B}{2} \\ 0 & \text{else.} \end{cases} \tag{6.7}$$

Certainly, this low-pass signal is not practical for radio transmission. However, extension of this analysis to bandpass transmit pulses is straightforward and leads to the same result, but expressions become lengthy then. This can also be seen by using an equivalent baseband model for the channel.

6.4.3 Estimation Bias and Variance

Given the path delays of the multipath components are uniformly i.i.d. within the pulse repetition period and the path gains are normally i.i.d. with zero mean and variance σ_α^2, i.e.

$$\gamma_n \sim \mathcal{U}(0, T_p) \text{ and } \alpha_n \sim \mathcal{N}(0, \sigma_\alpha^2)$$

for $n = 1, \ldots, M_p$ and the LOS-path corresponds to the parameter doublet with index 0, i.e.

$$\gamma_0 = 0 \text{ and } \alpha_0 = \text{const.},$$

then the estimation bias is given by

$$\mathsf{E}_{\alpha,\gamma}[\hat{\varepsilon}] \approx -\frac{T_p}{2\pi} \arg \mathsf{E}[z(t)] = \varepsilon \mod T_p$$

and the expected estimation error by

$$\mathsf{Var}_{\alpha,\gamma}[\hat{\varepsilon}] \approx \left(\frac{T_p}{2\pi}\right)^2 \frac{M_p}{\alpha_0^4} \left(\frac{3}{2}\sigma_\alpha^4 + \frac{(M_p-1)\sigma_\alpha^4}{BT_p} + \frac{2\alpha_0^2\sigma_\alpha^2}{(BT_p)^2}\right). \tag{6.8}$$

This shows that the estimator is approximately unbiased for the considered channel model. Additionally, an approximation of the estimation error is given, which shows the impact of signaling bandwidth. The approximation holds for a high LOS to multipath ratio, i.e. $\alpha_0^2 \gg \sigma_\alpha^2$, and if either the pulses are unmodulated or do not overlap (6.4).

This is derived as follows: Substituting the spectrum of the low-pass base pulse (6.7) into

(6.5) yields at the output of downconversion

$$z(t) = G\left(\tfrac{1}{T_p}\right) \frac{1}{T_p} \int\limits_{-\infty}^{\infty} e^{-j2\pi\nu\varepsilon} H(\nu) \tilde{P}(\nu) e^{-j2\pi\frac{\varepsilon}{T_p}+j2\pi\nu\varepsilon} H\left(\tfrac{1}{T_p}-\nu\right) \tilde{P}\left(\tfrac{1}{T_p}-\nu\right) d\nu$$

$$= G\left(\tfrac{1}{T_p}\right) \frac{1}{T_p B} e^{-j2\pi\frac{\varepsilon}{T_p}} \int\limits_{-\frac{B}{2}+\frac{1}{T_p}}^{\frac{B}{2}} H(\nu) H\left(\tfrac{1}{T_p}-\nu\right) d\nu, \qquad (6.9)$$

since

$$\tilde{P}(\nu)\tilde{P}\left(\tfrac{1}{T_p}-\nu\right) = \begin{cases} \frac{1}{B} & \text{for } -\frac{B}{2}+\frac{1}{T_p} < \nu < \frac{B}{2} \\ 0 & \text{else.} \end{cases}$$

The integral in (6.9) can be written as a sum, if the signal $r(t)$ at the input of the squarer is periodic. Motivation for this assumption is that for a rather low bandwidth it is reasonable to transmit unmodulated pulses (narrowband case), i.e. $\tilde{a}_k = \text{const.}$ For a high bandwidth (UWB), the overlapping of the pulses (intersymbol interference) is negligible. For both cases the following approximation holds with equality

$$\int\limits_{-\frac{B}{2}+\frac{1}{T_p}}^{\frac{B}{2}} H(\nu) H\left(\tfrac{1}{T_p}-\nu\right) d\nu \approx \frac{1}{T_p} \sum_{l=-\left\lceil\frac{T_p B}{2}\right\rceil+1}^{\left\lfloor\frac{T_p B}{2}\right\rfloor} H\left(\tfrac{l}{T_p}\right) H\left(\tfrac{1-l}{T_p}\right).$$

To get rid of the floor and ceil operation in the index boundaries and to simplify notation, we assume (without loss of generality) $\frac{T_p B}{2} \in \mathbb{N}$. Substituting now the spectrum of the channel impulse response (6.6) leads to

$$z(t) \approx G\left(\tfrac{1}{T_p}\right) \frac{1}{T_p^2 B} e^{-j2\pi\frac{\varepsilon}{T_p}} \sum_{n=0}^{M_p} \sum_{m=0}^{M_p} \alpha_n\alpha_m e^{-j2\pi\frac{1}{T_p}\gamma_m} \sum_{l=-\frac{T_p B}{2}+1}^{\frac{T_p B}{2}} e^{-j2\pi\frac{1}{T_p}(\gamma_n-\gamma_m)l}.$$

The sum over l in the last term can be simplified with the identity

$$\sum_{l=-\frac{T_p B}{2}+1}^{\frac{T_p B}{2}} e^{-j2\pi\frac{1}{T_p}(\gamma_n-\gamma_m)l}$$

$$= \begin{cases} T_p B & \text{for } \gamma_n-\gamma_m = kT_p, \; \forall k \in \mathbb{Z} \\ e^{-j\pi\frac{1}{T_p}(\gamma_n-\gamma_m)} \dfrac{\sin(\pi(\gamma_n-\gamma_m)B)}{\sin\left(\pi(\gamma_n-\gamma_m)\frac{1}{T_p}\right)} & \text{else.} \end{cases}$$

The output $z(t)$ can then be written compactly in matrix notation:

$$z(t) = \boldsymbol{\alpha}^T \boldsymbol{\Gamma} \boldsymbol{\alpha} \cdot e^{-j2\pi\frac{\epsilon}{T_p}}, \tag{6.10}$$

where $\boldsymbol{\alpha} = [\alpha_0, \ldots, \alpha_{M_p}]^T$ and

$$\boldsymbol{\Gamma} = \begin{bmatrix} 1 & \Gamma_{0,1} & \cdots & \Gamma_{0,M_p} \\ \Gamma_{1,0} & e^{-j2\pi\frac{\gamma_1}{T_p}} & \ddots & \vdots \\ \vdots & \ddots & \ddots & \Gamma_{(M_p-1),M_p} \\ \Gamma_{M_p,0} & \cdots & \Gamma_{M_p,(M_p-1)} & e^{-j2\pi\frac{\gamma_{M_p}}{T_p}} \end{bmatrix} \frac{G(\frac{1}{T_p})}{T_p}$$

with

$$\Gamma_{n,m} = \frac{1}{T_p B} e^{-j\pi\frac{1}{T_p}(\gamma_n+\gamma_m)} \frac{\sin\left(\pi(\gamma_n - \gamma_m)B\right)}{\sin\left(\pi(\gamma_n - \gamma_m)\frac{1}{T_p}\right)}.$$

Estimation Bias: Linearization of $\arg(\cdot)$ yields for small variances

$$\mathsf{E}[\arg z(t)] \approx \arg \mathsf{E}[z(t)].$$

Using (6.10), the phase of the expectation of the downconversion output $z(t)$ can be written as

$$\arg \mathsf{E}[z(t)] = \left(\arg e^{-j2\pi\frac{\epsilon}{T_p}} + \arg \mathsf{E}_{\alpha,\gamma}\left[\boldsymbol{\alpha}^T \boldsymbol{\Gamma} \boldsymbol{\alpha}\right]\right) \bmod 2\pi.$$

The expectation of the quadratic form yields

$$\begin{aligned}
\mathsf{E}_{\alpha,\gamma}\left[\boldsymbol{\alpha}^T \boldsymbol{\Gamma} \boldsymbol{\alpha}\right] &= \mathsf{E}_{\gamma}\left[\mathsf{E}_{\alpha}\left[\boldsymbol{\alpha}^T \boldsymbol{\Gamma} \boldsymbol{\alpha}\right]\right] \\
&= \mathsf{E}_{\gamma}\left[\mathsf{E}_{\alpha}\left[\mathrm{Tr}\left(\boldsymbol{\alpha}^T \boldsymbol{\Gamma} \boldsymbol{\alpha}\right)\right]\right] \\
&= \mathsf{E}_{\gamma}\left[\mathsf{E}_{\alpha}\left[\mathrm{Tr}\left(\boldsymbol{\Gamma} \boldsymbol{\alpha} \boldsymbol{\alpha}^T\right)\right]\right] \\
&= \mathsf{E}_{\gamma}\left[\mathrm{Tr}\left(\boldsymbol{\Gamma} \mathsf{E}_{\alpha}\left[\boldsymbol{\alpha} \boldsymbol{\alpha}^T\right]\right)\right] \\
&= \mathsf{E}_{\gamma}\left[\mathrm{Tr}\left(\boldsymbol{\Gamma} \mathrm{Cov}[\boldsymbol{\alpha}] + \boldsymbol{\Gamma}\mathsf{E}[\boldsymbol{\alpha}]\mathsf{E}[\boldsymbol{\alpha}]^T\right)\right] \\
&= \mathsf{E}_{\gamma}\left[\mathrm{Tr}\left(\boldsymbol{\Gamma} \mathrm{Cov}[\boldsymbol{\alpha}]\right) + \mathsf{E}[\boldsymbol{\alpha}]^T \boldsymbol{\Gamma} \mathsf{E}[\boldsymbol{\alpha}]\right] \\
&= \mathrm{Tr}\left(\mathsf{E}_{\gamma}[\boldsymbol{\Gamma}] \mathrm{Cov}[\boldsymbol{\alpha}]\right) + \mathsf{E}[\boldsymbol{\alpha}]^T \mathsf{E}_{\gamma}[\boldsymbol{\Gamma}] \mathsf{E}[\boldsymbol{\alpha}],
\end{aligned} \tag{6.11}$$

where we used $E[\mathrm{Tr}(\mathbf{A})] = \mathrm{Tr}(E[\mathbf{A}])$ and $\mathrm{Tr}(\mathbf{ABC}) = \mathrm{Tr}(\mathbf{CAB})$. Further, we have

$$E[\boldsymbol{\alpha}] = [\alpha_0, 0, \ldots, 0]^T \tag{6.12}$$

and

$$\mathrm{Cov}[\boldsymbol{\alpha}] = \mathrm{diag}\left([0, \sigma_\alpha^2, \ldots, \sigma_\alpha^2]^T\right). \tag{6.13}$$

The expectation of $\boldsymbol{\Gamma}$ with respect to γ yields

$$E_\gamma[\boldsymbol{\Gamma}] = \mathrm{diag}\left([1, 0, \ldots, 0]^T\right) \frac{G\left(\frac{1}{T_p}\right)}{T_p}.$$

Substituting this, the value of (6.11) then computes to

$$E_{\alpha,\gamma}\left[\boldsymbol{\alpha}^T \boldsymbol{\Gamma} \boldsymbol{\alpha}\right] = \alpha_0^2 \frac{G\left(\frac{1}{T_p}\right)}{T_p}.$$

Since it is assumed that the transfer function of the post detection filter $g(t)$ is real, it follows

$$\arg E_{\alpha,\gamma}[\boldsymbol{\alpha}^T \boldsymbol{\Gamma} \boldsymbol{\alpha}] = 0$$

and further

$$\arg E[z(t)] = \frac{2\pi}{T_p}(\varepsilon \mod T_p).$$

Thus, the estimator is approximately unbiased for this channel model.

Error Variance: Applying again linearization of $\arg(\cdot)$ yields for $\alpha_0^2 \gg \sigma_\alpha^2$ for the expected estimation error

$$\begin{aligned}
\mathrm{Var}_{\alpha,\gamma}[\hat{\varepsilon}] &= \left(\tfrac{T_p}{2\pi}\right)^2 \mathrm{Var}[\arg \boldsymbol{\alpha}^T \boldsymbol{\Gamma} \boldsymbol{\alpha}] \\
&\approx \left(\tfrac{T_p}{2\pi}\right)^2 \frac{E\left[\mathrm{Im}\left\{\boldsymbol{\alpha}^T \boldsymbol{\Gamma} \boldsymbol{\alpha}\right\}^2\right]}{|E[\boldsymbol{\alpha}^T \boldsymbol{\Gamma} \boldsymbol{\alpha}]|^2}.
\end{aligned}$$

The second moment of the imaginary part of the quadratic form $\alpha^T \Gamma \alpha$ can be written as

$$
\begin{aligned}
\mathsf{E}_{\alpha,\gamma}\left[\operatorname{Im}\left\{\alpha^T \Gamma \alpha\right\}^2\right] &= \mathsf{E}_{\alpha,\gamma}\left[\left(\alpha^T \operatorname{Im}\{\Gamma\}\,\alpha\right)^2\right] \\
&= \mathsf{E}_\gamma\left[\operatorname{Var}_\alpha\left[\alpha^T \operatorname{Im}\{\Gamma\}\,\alpha\right]\right] \\
&= \mathsf{E}_\gamma\left[2\operatorname{Tr}\left[(\operatorname{Im}\{\Gamma\}\operatorname{Cov}[\alpha])^2\right]\right] \\
&\quad + \mathsf{E}_\gamma\left[4\mathsf{E}[\alpha]^T \operatorname{Im}\{\Gamma\}\operatorname{Cov}[\alpha]\operatorname{Im}\{\Gamma\}\mathsf{E}[\alpha]\right],
\end{aligned}
$$

where we used Theorem 5.2c from [92] to obtain the variance of the quadratic form[3]. With (6.12) and (6.13) follows

$$
\begin{aligned}
\mathsf{E}_{\alpha,\gamma}\left[\operatorname{Im}\left\{\alpha^T \Gamma \alpha\right\}^2\right] &= 2\sigma_\alpha^4 \sum_{n=1}^{M_p}\sum_{m=1}^{M_p}\mathsf{E}_\gamma\left[\operatorname{Im}\{\Gamma_{n,m}\}^2\right] + \sum_{n=1}^{M_p}\mathsf{E}_\gamma\left[\operatorname{Im}\{\Gamma_{n,n}\}^2\right] + \\
&\quad + 4\alpha_0^2\sigma_\alpha^2 \sum_{m=1}^{M_p}\mathsf{E}_\gamma\left[\operatorname{Im}\{\Gamma_{0,m}\}^2\right].
\end{aligned}
$$

The expectation with respect to γ computes to:

i) For $n = m$:

$$
\mathsf{E}_\gamma\left[\operatorname{Im}\{\Gamma_{n,n}\}^2\right] = \mathsf{E}_\gamma\left[\left(\frac{G\left(\frac{1}{T_p}\right)}{T_p}\sin\left(2\pi\frac{\gamma_n}{T_p}\right)\right)^2\right] = \frac{G^2(\frac{1}{T_p})}{2T_p^2} \tag{6.14}
$$

ii) For $n \neq m$:

$$
\begin{aligned}
\mathsf{E}_\gamma\left[\operatorname{Im}\{\Gamma_{n,m}\}^2\right] &= \mathsf{E}_\gamma\left[\left(\frac{G\left(\frac{1}{T_p}\right)}{T_p^2 B}\frac{\sin(\pi(\gamma_n+\gamma_m)\frac{1}{T_p})\sin\left(\pi(\gamma_n-\gamma_m)B\right)}{\sin\left(\pi(\gamma_n-\gamma_m)\frac{1}{T_p}\right)}\right)^2\right] \\
&= \frac{G^2(\frac{1}{T_p})}{2T_p^3 B} \tag{6.15}
\end{aligned}
$$

iii) For $n = 0 \neq m$:

$$
\mathsf{E}_\gamma\left[\operatorname{Im}\{\Gamma_{0,m}\}^2\right] = \mathsf{E}_\gamma\left[\left(\frac{G\left(\frac{1}{T_p}\right)}{T_p^2 B}\sin\left(\pi\gamma_m B\right)\right)^2\right] = \frac{G^2(\frac{1}{T_p})}{2T_p^4 B^2} \tag{6.16}
$$

---

[3]If $\mathbf{y}$ is $\mathcal{N}(\mu,\Sigma)$, then

$$
\operatorname{Var}\left[\mathbf{y}^T \mathbf{A}\mathbf{y}\right] = 2\operatorname{Tr}\left[(\mathbf{A}\Sigma)^2\right] + 4\mu^T \mathbf{A}\Sigma\mathbf{A}\mu.
$$

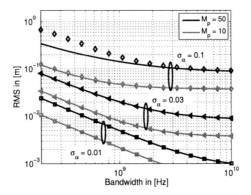

Fig. 6.4: Simulation (markers) and approximation (lines) of RMS of ranging error vs. bandwidth for channel model ($\frac{1}{T_p} = 50$ MHz).

The derivation of the integral expressions to evaluate the expectations is given in Appendix A. Collecting terms yields

$$\frac{\mathsf{E}\left[\operatorname{Im}\left\{\boldsymbol{\alpha}^T\boldsymbol{\Gamma}\boldsymbol{\alpha}\right\}^2\right]}{|\mathsf{E}\left[\boldsymbol{\alpha}^T\boldsymbol{\Gamma}\boldsymbol{\alpha}\right]|^2} = \frac{M_p}{\alpha_0^4}\left(\frac{3}{2}\sigma_\alpha^4 + \frac{(M_p-1)\sigma_\alpha^4}{BT_p} + \frac{2\alpha_0^2\sigma_\alpha^2}{(BT_p)^2}\right)$$

and finally leads to (6.8).

Expression (6.8) shows the influence of the bandwidth of the transmit pulse on the estimation accuracy. The higher the bandwidth of the transmit pulse, the better becomes the ranging accuracy. However, there is an error floor that is determined by the term $\frac{3}{2}\sigma_\alpha^4$, which is independent of B. The second term scales with $\frac{1}{BT_p}$ and corresponds to overlap of the multipath terms. The higher the bandwidth, the shorter are the pulses and therefore the probability of overlap decreases. The third term scales with $\frac{1}{(BT_p)^2}$ and is related to the overlap of multipath components with the LOS path.

6.5 Performance Results

This section presents the performance analysis of the proposed ToA estimator. First, the estimation error according to (6.8) is compared to Monte Carlo simulation. Second, the estimation error and bias is evaluated for a rich multipath indoor environment based on measured

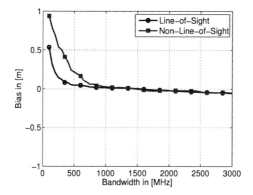

Fig. 6.5: Bias of ranging estimate vs. bandwidth for measured channels.

channel impulse responses. The pulse repetition frequency has been chosen to $\frac{1}{T_p} = 50$ MHz, i.e. the uncertainty window is 6 m. This would be an adequate value e.g. for motion tracking in an indoor environment. As figure of merit, we consider the root mean square of the estimation error.

Fig. 6.4 depicts the influence of the bandwidth of the transmit pulse according to (6.8). The RMS is plotted for different values of σ_α and number of scatterers M_p. The magnitude of the direct LOS path is set to $\alpha_0 = 1$. The solid lines correspond to the analytical result and show very good matching compared to the Monte Carlo simulation (markers) for sufficiently small σ_α. This validates the approximations in Section 6.4 in the derivation of the expected estimation error. The result shows how the positioning accuracy increases with increasing bandwidth. In particular, the $\frac{1}{B}$-scaling of the estimation error can be seen in Fig. 6.4 for the small value of σ_α. However, the estimation accuracy saturates at a certain error floor, which is given by the bandwidth independent term in (6.8).

For the performance evaluation based on measured channel impulse responses two different scenarios are chosen: a typical LOS situation and a non-LOS situation in an indoor environment with rich multipath propagation. Fig. 6.5 shows the estimation bias versus bandwidth and Fig. 6.6 the RMS based on 8820 LOS measurements and 5020 non-LOS measurements for bandpass signaling with center frequency $f_c = 4.5$ GHz. We used the same measurement setup and postprocessing as in Section 3.5.2 with the floor plan shown in Fig. 3.14 on Page 47. The regions 1-14 are combined to the LOS case and the regions 15-22 to the non-LOS scenario.

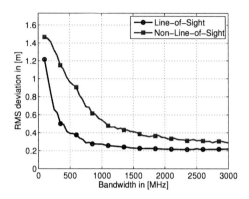

Fig. 6.6: RMS of ranging error vs. bandwidth for measured channels.

The results show that the estimation bias is small for a sufficient bandwidth. For the LOS situation, the positioning error saturates at about 20 cm compared to 30 cm for the non-LOS case. However, for a low signaling bandwidth, the estimation performance is very poor. The upper limit of the y-axis in Fig. 6.6 is set to $\sqrt{3}$ m ≈ 1.73 m, which would correspond to the RMS of uniformly distributed guessing in the uncertainty window of 6 m. This shows that in this environment the phase of narrowband signals is almost useless for positioning and narrowband systems would not give a reliable position estimate. In contrast, when using UWB pulses with a bandwidth of 2 GHz or more the systems performance is improved substantially.

6.6 Summary

Low complexity position estimation based on UWB energy detection and spectral estimation has been presented and analyzed. Moreover, the influence of the bandwidth is investigated and a closed form approximation of the estimation variance is presented. Performance evaluation based on measured channel impulse responses show that up to 20 cm accuracy can be reached.

Chapter 7

Radar Imaging based Multipath Delay Prediction

Communication and localization in UWB sensor networks can strongly benefit from a priori channel knowledge. The communication can efficiently be optimized based on the location of the nodes to save channel estimation overhead and feedback. The position estimation can be improved when the distribution of the channel is known. However, both require a database with the mapping from locations to channel statistics. In the following two chapters, which are the third part of this thesis, we present and verify an efficient way to obtain this database.

In this chapter[1], we propose a radar imaging based approach to predict the channel for an arbitrary position. This is done in three steps: First, a radar image of the environment is created using measured training data. The training data can easily be obtained from channel estimates in a UWB sensor network with mobile nodes. The radar image contains the strong reflectors and scatterers in the environment. Specifically, a scattering coefficient map is generated, where the large synthetic aperture of the distributed and moving antennas is used. Second, the map is used to reconstruct path gains and path delays. Thus, the channel response is predicted for arbitrary transmitter and receiver positions. Finally, dominant multipath delays are extracted to suppress reconstruction errors and out-of-band noise. The proposed algorithm is validated by anechoic chamber measurements with controlled reflectors.

7.1 Introduction

Radar imaging enables a UWB sensor network to sense the environment. Every receive signal contains information about the surrounding area and the topology. Due to the high

---
[1]Parts of this chapter have been published in [45].

bandwidth, the multipath components from strong scatterers or reflectors can be resolved with high accuracy. With a high density of nodes or with mobile nodes, a large amount of channel information can be collected. This allows to create an image of the surrounding area and to identify and locate objects that influence the radio propagation.

The information about the surrounding area enables pioneering applications of UWB sensor networks. Before we present the contribution of this chapter, we give a short overview on selected UWB imaging techniques. For a comprehensive literature overview, see [26] and the references therein. Generally, the imaging approaches can be divided into two groups, *tomographic* and *radar-based* systems.

In *tomographic imaging*, the object of interest is surrounded by sensors. The goal is to see inside this object and to reconstruct its internal properties. UWB tomographic imaging aims to obtain slices of the object of interest without destroying it. UWB imaging is a nondestructive testing method and for humans less harmful than x-ray scans. This leads to many promising applications in industry, medical, civil and security areas. A prominent example for UWB tomographic radar imaging is breast cancer detection, see [27].

The goal of *radar-based* systems is to determine the shape and location of an object. Usually, radar systems use directional antennas or large antenna arrays to scan the area of interest. In synthetic aperture radar (SAR) the echo waveforms are measured successively from different positions. A well-known application of UWB radar is through-wall imaging for disaster recovery. After a building collapse, the UWB radar can help to find trapped humans or cavities. In literature, it has been shown that human respiratory and heartbeat activity can be detected from the channel information [30, 93, 94]. Successful identification could be demonstrated through 30 cm thick walls. Other potential applications are vital signs surveillance systems for elderly people. The advantage of UWB systems is the preserved privacy compared to video systems.

In [95], a UWB radar-based human fall detection system is presented. It is shown that in a UWB body area network we can benefit from the multipath propagation and actually use reflectors and scatterers for human movement analysis and posture detection. Fall detection systems could save lives with an automatic emergency call in case of an accident, e.g. for elderly people or athletes. Specifically, wireless nodes attached to the body are considered. An algorithm is developed to track the movement of the human body and to perform fall detection. Assuming that the channel impulse response is measured successively in small time steps, the evolution of the multipath components give information about the movement of the person. To track the multipath delays, state space estimation is used and the corresponding Particle Filter and Kalman Filter are developed. Pattern recognition applied to the delay trajectories make the detection of a fall possible. Simulations and real world measurements prove that the developed algorithms can detect a fall successfully, making it a promising secondary application of UWB body area networks.

When considering a UWB sensor network, the goal of imaging is to obtain a map of the environment with the scatterers and reflectors. The reconstruction algorithm generates the image from the measured channels and node positions. Standard reconstruction algorithms are known from seismic engineering and geophysics [29, 96]. The reconstruction problem was tackled in the early 1980's to explore the earth's subsurface by reflected seismic waves. Much research has been conducted in this field due to the lucrative business of in-ground resource exploration. Reflection seismology can be used to find and locate oil, coal or mineral reservoirs. The well-known reconstruction algorithms are derived from the wave equations and Huygen's principle. Kirchhoff migration [97] is the most basic imaging principle. The reconstruction formula is given by

$$U(x,y,z,0) = -\frac{1}{2\pi}\frac{\partial}{\partial z}\int\int \frac{U\left(x_0,y_0,0,\frac{r}{c}\right)}{r}\mathrm{d}x_0\mathrm{d}y_0,$$

where $U(x,y,z,t)$ denotes the scalar wave field, c the propagation speed, and

$$r = \sqrt{(x-x_0)^2 + (y-y_0)^2 + z^2}.$$

With coincident transmitter and receiver the wave field holds $U(x_0,y_0,0,t) = G(x_0,y_0,0,2t)$, where $G(x_0,y_0,0,t)$ is the measured field data (impulse response) at position $[x_0,y_0]^T$. Literature proposes many UWB imaging approaches based on or modified from Kirchhoff migration, see e.g. [28, 98–100].

In [101], different UWB imaging reconstruction algorithms are compared and demonstrated. The framework is based on ray-optical propagation of the signals between transmitting and receiving antennas. Therefore, the channel impulse response for arbitrary environments can be generated by ray-tracing. A simulation toolbox is developed, including a graphical user interface to investigate the effect of the environment, node density and reconstruction scheme. The software allows to construct different scenarios and to compare the model with the reconstruction. This allows to evaluate the standard imaging algorithms and to get an intuition on the requirements and characteristics of UWB radar imaging.

In this thesis, we introduce UWB imaging as a method to predict the channel for arbitrary positions in a stationary environment from a set of training samples. The UWB channel depends on the surrounding objects and on the position of transmitter and receiver. The principle of the multipath delay prediction is based on the identification and localization of strong scatterers and reflectors. Once they are known, the expected multipath delay for arbitrary transmitter and receiver positions can be determined from the geometry. Specifically, multipath delay prediction makes use of UWB radar imaging techniques with distributed transmit and receive antennas and is based on a three step approach.

In the first step, a synthetic aperture radar image of the environment is generated based on training data. We use channel estimates measured at different receiver and transmitter positions as training data, which can be obtained from a UWB communication network with mobile nodes without additional costs. We propose to use a low complexity imaging algorithm to create a map of the scattering coefficients for the area of interest. This map gives information about the location and shape of strong reflectors and scatterers. The second step enables reconstruction of channel responses for arbitrary transmitter and receiver positions. The predicted channel response of new positions is synthesized from the radar image. Finally, we use WRELAX [83] to extract the dominant multipath delays of the predicted channel response. The contribution of this chapter can be summarized as follows:

- We provide an algorithm to predict multipath delays for arbitrary transmitter and receiver positions.

- We exploit synergies of UWB imaging, localization and communication. Sensing of the environment is proposed to improve position estimation by channel response reconstruction and *a priori* channel information for communication.

- The imaging and reconstruction algorithm is validated with anechoic chamber measurements with controlled reflectors.

The prediction of the channel based on the radar images has a number of very promising applications. The *a priori* channel information can be used for localization as well as communication. When considering geometrical localization techniques, such as ToA or AoA, the knowledge of the multipath can increase the position estimation accuracy to a great extent. In fact, the performance of ToA estimation degrades tremendously in harsh propagation environments. The reason for this is that the LOS path may not be distinguishable from multipath or may even not exist. This leads to high position estimation errors and biased results, when no *a priori* channel knowledge is available. Considering the Cramér-Rao lower bound for the unbiased position estimate as presented in [102], we observe that the squared position error even diverges when the separation of overlapping paths goes to zero. However, if channel information is available, the position estimation can profit from multipath components. An example for this is UWB position estimation with floor plan information [103]. It requires a priori knowledge of the propagation environment and the presence of strong reflectors.

Considering localization based on fingerprinting, UWB radar imaging based multipath delay prediction can be used to interpolate between training samples. Generally, the performance of location fingerprinting techniques depends on the accuracy of the training data. The positioning error is usually in the order of the distances of the training samples. Therefore, for high definition positioning, the number of required training samples grows large, which is expensive and impractical to obtain. In [39], location fingerprinting with UWB channel impulse responses has been introduced and investigated in detail. Several methods to obtain the *a priori* distribution of the channel have been proposed. One approach is to select a parameterized distribution from a set of well-known distributions. In [39], this is done by Akaike's information criterion and the parameters of the distribution are estimated according to the maximum likelihood. A promising approach is to assume Gaussian channel taps and to characterize the channel by the mean and covariance matrix. However, to obtain accurate parameter estimates, a large set of training samples is necessary. Therefore, [39] extents the parameter estimation with an efficient training phase. This method is based on a first order approximation of the path delays and path gains, see [104] for details. Thus, the expected delays and gains can be predicted for a small area. However, this still requires a large number of training samples and only moderate accuracy can be obtained. Although this method takes the geometry of transmitter and receiver into account, it does not consider the location and shape of reflectors and scatterers.

We present a method to obtain the *a priori* knowledge of multipath delays, which is based on a radar image of the environment. Knowledge of multipath delays is of great benefit for geometric as well as fingerprinting based position estimation approaches. Considering ToA estimation, it helps to identify the LOS path and makes position estimation robust against multipath. For fingerprinting methods, multipath prediction can be used to efficiently interpolate between training samples and thus increase accuracy substantially.

The remainder of the chapter is structured as follows. Section 7.2 describes the system model and in Section 7.3 the imaging method to generate the scattering coefficient map is presented. The algorithm for channel response prediction is introduced in Section 7.4 and conclusions are drawn in Section 7.6.

7.2 System Setup and Problem Formulation

As in position estimation, we consider a UWB sensor network with anchor nodes and agent nodes. The anchor nodes are the cluster heads, i.e. stationary and their absolute position is known with high accuracy. The agents are the sensor nodes, i.e. mobile and usually subject to position estimation. Here, we want to predict the multipath conditions for arbitrary agent positions.

Considering radio transmission from agent to anchors, the receive signal is modeled by

$$r(t) = \sum_{n=0}^{M_p} \alpha_n s(t - \tau_n) + w(t),$$

where $s(t)$ and $w(t)$ denote transmit signal and noise, respectively. The smallest path delay τ_n corresponds to the LOS path and dominant propagation paths are characterized by large path gains α_n. We denote the receive signal correlated with the transmit signal $s(t)$ as channel response $\psi(t)$, where $s(t)$ preferably has a flat spectrum in the considered bandwidth.[2]

The channel response depends on the position of the transmitting agent as well as on the position of the receiving anchor. Suppose an agent is moved though the coverage area of the considered network and the different anchors successively record channel responses. In combination with the position information of the anchor and agents, this forms a training data set. The training data set $\mathcal{M}$ is defined as

$$\mathcal{M} = \left\{ \psi^{(1)}(t), \mathbf{p}_{SN}^{(1)}, \mathbf{p}_A^{(1)}, \dots, \psi^{(N)}(t), \mathbf{p}_{SN}^{(N)}, \mathbf{p}_A^{(N)} \right\},$$

where $\psi^{(k)}(t)$, $\mathbf{p}_{SN}^{(k)} = [x_{SN}^{(k)}, y_{SN}^{(k)}]^T$, $\mathbf{p}_A^{(k)} = [x_A^{(k)}, y_A^{(k)}]^T$ denote channel response, transmitter position and receiver position for measurement number $k = 1, \dots, N$, respectively. Given the data set $\mathcal{M}$, we want to find the delays of the dominant paths for an arbitrary agent position $\mathbf{p}_{SN}$ and anchor position $\mathbf{p}_A$. We assume a static environment and a constant velocity medium with propagation speed $c = c_0$, the speed of light in vacuum.

7.3 Scattering Coefficient Map of Environment

To generate an image of the environment we use synthetic aperture processing. We estimate the scattering coefficient $\hat{\alpha}_{\mathcal{M}}(\mathbf{r})$ for every point $\mathbf{r}$ in the coverage area of the considered network. The reconstruction algorithm is given by

$$\hat{\alpha}_{\mathcal{M}}(\mathbf{r}) = \frac{1}{N} \sum_{k=1}^{N} R_1^{(k)}(\mathbf{r}) R_2^{(k)}(\mathbf{r}) \cdot \psi^{(k)} \left(\frac{R_1^{(k)}(\mathbf{r}) + R_2^{(k)}(\mathbf{r})}{c} \right), \tag{7.1}$$

---

[2]Without noise and with $s(t)$ as bandpass pulse, the channel response corresponds to the bandpass-filtered channel impulse response, i.e. $\psi(t) = \tilde{h}(t)$.

where $R_1^{(k)}(\mathbf{r})$ and $R_2^{(k)}(\mathbf{r})$ correspond to the distance between point $\mathbf{r}$ and agent and anchor, respectively. In the following, we consider the 2D case, i.e. $\mathbf{r} = [x, y]^T$. The distances are then given by

$$R_1^{(k)}(x, y) = \sqrt{\left(x - x_{\mathrm{SN}}^{(k)}\right)^2 + \left(y - y_{\mathrm{SN}}^{(k)}\right)^2}$$

and

$$R_2^{(k)}(x, y) = \sqrt{\left(x - x_{\mathrm{A}}^{(k)}\right)^2 + \left(y - y_{\mathrm{A}}^{(k)}\right)^2}.$$

The principle of the imaging algorithm is that each point $\mathbf{r}_1 = [x_1, y_1]^T$ in the coverage area can be mapped to a delay in the channel responses. This delay depends on the agent and anchor positions and is determined by $\frac{1}{c}(R_1^{(k)}(\mathbf{r}_1) + R_2^{(k)}(\mathbf{r}_1))$. If we average now over many agent positions, path gains add up coherently in case of a reflector or scatterer and average out, if no object is present. To compensate the effect of path-loss, the channel response is scaled by the distances of the considered point to the transmitter and receiver. Hence, we asymptotically obtain a map of intensity and location of reflectors and scatterers. Note that the LOS component in the channel response does not correspond to a reflector or scatterer. Therefore, we substitute the LOS part of the channel responses $\psi^{(k)}(t)$ by zeros before generating the scattering coefficient map according to (7.1).

In geophysics and seismology, equation (7.1) is known as diffraction summation migration [96]. It is straightforward to extend the presented multipath delay prediction algorithm by other imaging techniques, such as Kirchhoff migration or finite difference methods for wave propagation and migration [29]. However, we omit this due to computational complexity, the distributed transmit and receive antenna and non-uniform measurement steps.

Fig. 7.1 shows the result of the imaging algorithm for a controlled environment in an anechoic chamber with two metallic reflectors. A picture of the setup is given in Fig. 7.2. The left hand side of Fig. 7.1 depicts the position of reflectors, anchor and agent. Channel responses have been measured in the bandwidth from 2 to 6 GHz, while the agent was moved on a horizontal 1 cm grid over an area of 0.28×0.28 cm. The measurement setup and successive postprocessing is described in detail in [39]. The right hand side of Fig. 7.1 shows the magnitude of the scattering coefficient map according to (7.1) based on $N = 840$ channel responses. The position and intensity of the two reflectors can be clearly determined.

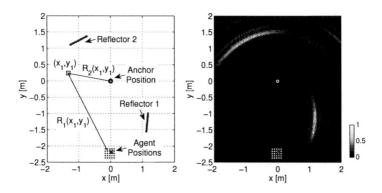

Fig. 7.1: Anechoic chamber measurements with two reflectors: setup (left) and magnitude of scattering coefficient map (right)

Fig. 7.2: Picture of anechoic chamber measurements with controlled reflectors (from [39]).

7.4 Channel Response Prediction

So far, we have extrapolated the time domain channel responses to the spatial dimension by generating a scattering coefficient map. Next, the spatial data is transformed to a channel response. The knowledge of the propagation environment enables to predict the channel response for an arbitrary position. This is done by synthesizing the channel response from the scattering coefficients with the corresponding delays. The predicted channel response $\hat{\Psi}_{\mathrm{p_{SN},p_A}}(f)$ in frequency domain with new transmitter and receiver position $\mathrm{p_{SN}} = [x_{\mathrm{SN}}, y_{\mathrm{SN}}]^T$ and $\mathrm{p_A} = [x_{\mathrm{A}}, y_{\mathrm{A}}]^T$ is given by:

$$\hat{\Psi}_{\mathrm{p_{SN},p_A}}(f) = \int_x \int_y \frac{\hat{\alpha}_{\mathcal{M}}(x,y)e^{-j2\pi f\frac{R_1(x,y)+R_2(x,y)}{c}}}{R_1(x,y)R_2(x,y)}\,\mathrm{d}x\mathrm{d}y, \tag{7.2}$$

where $R_1(x,y)$ and $R_2(x,y)$ are the distances between $[x,y]^T$ and $\mathrm{p_{SN}}$ and $\mathrm{p_A}$, respectively. The delay of a scatterer at position $[x,y]^T$ with intensity $\hat{\alpha}_{\mathcal{M}}(x,y)$ leads to a phase shift depending on the distance to the receiver and transmitter position. The path loss is incorporated by appropriate scaling with the distance.

Finally, the dominant multipath components are extracted from $\hat{\Psi}_{\mathrm{p_{SN},p_A}}(f)$. This gives an estimate of the multipath delays as a function of the agent and anchor position. The $\tilde{M}_p$-strongest predicted multipath delays $\hat{\tau}_l$ are obtained by

$$(\hat{\tau}_1,\ldots,\hat{\tau}_{\tilde{M}_p}) = \underset{\substack{\tilde{\alpha}_1,\ldots,\tilde{\alpha}_{\tilde{M}_p} \\ \tilde{\tau}_1,\ldots,\tilde{\tau}_{\tilde{M}_p}}}{\arg\min} \int_{-\infty}^{\infty} \left| \hat{\Psi}_{\mathrm{p_{SN},p_A}}(f) - |S(f)|^2 \sum_{n=1}^{\tilde{M}_p} \tilde{\alpha}_l e^{-j2\pi f\tilde{\tau}_l} \right|^2 \mathrm{d}f,$$

where $S(f)$ denotes the Fourier transform of the transmit signal $s(t)$. We propose to solve the optimization problem with WRELAX, see [83].

For the sake of computational complexity, we introduce sampling of the scattering coefficient map with spacing Δx and Δy and of the channel response with a sampling frequency f_s. In discrete time, the channel response prediction (7.2) can then be approximated by

$$\hat{\psi}_{\mathrm{p_{SN},p_A}}[k] = \sum_l \sum_m \frac{\hat{\alpha}_{\mathcal{M}}(x_l,y_m)\Delta x\Delta y}{R_1(x_l,y_m)R_2(x_l,y_m)} \cdot \delta\left[k - \frac{f_s}{c}(R_1(x_l,y_m)+R_2(x_l,y_m))\right]$$

where $\delta[\cdot]$ denotes the Kronecker delta with the argument rounded to the nearest integer.

Fig. 7.3 shows a reconstructed channel response from the controlled anechoic chamber environment. The channel response corresponds to agent position $[0.018\text{ m}, -2.177\text{ m}]^T$,

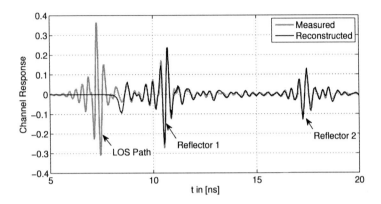

Fig. 7.3: Reconstructed and measured channel response of controlled anechoic chamber environment with two reflectors

which is marked by a red cross in Fig. 7.1. Additionally, the measured channel response at this position is plotted. Note that the measured channel response has not been used for the reconstruction, i.e. it was excluded from the training set to generate the scattering coefficient map. The reconstructed and the measured channel response fit very well. The reconstructed channel response does not contain the LOS path, since it does not originate from reflection or scattering.

7.5 Performance Evaluation of Multipath Prediction

To evaluate the accuracy of the multipath prediction, we compared all reconstructed with the measured channel impulse responses. For each position, the scattering coefficient map is generated without the considered measurement and the channel impulse response is reconstructed with the proposed algorithm. Subsequently, the two strongest multipath delays are extracted using WRELAX. The CDF of the path delay estimation error is shown in Fig. 7.4. The prediction error is evaluated in centimeter, i.e. 3 cm correspond to approximately 0.1 ns. It can be seen that the prediction error is in most cases much smaller than 1 cm. This is notable, since the measurements are taken on a 1 cm grid. Therefore, we conclude that the prediction algorithm can successfully interpolate between channel impulse responses and can be used to predict the channel. Note that the error floor with less than 5% of large estimation

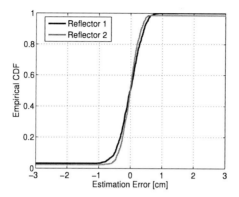

Fig. 7.4: CDF of path delay estimation error from anechoic chamber measurements

errors is due to a wrong allocation of the multipath delays. In those cases, the first reflector is detected twice due to a small distortion of the pulse. However, the second reflector would be found as the next multipath delay. Thus, the large error is only due to the pairing problem of the delays and the predicted channel impulse response has still a good quality.

This can also be observed when looking at the correlation between the measured and reconstructed channel impulse response. Fig. 7.5 plots the correlation coefficient for all positions of the $29 \times 29\,\text{cm}$ grid. The correlation coefficient is defined by

$$\rho_{\text{PSN,PA}} = \frac{\int_{-\infty}^{\infty} \hat{\psi}_{\text{PSN,PA}}(t) \cdot \psi_{\text{PSN,PA}}(t)\mathrm{d}t}{\sqrt{\int_{-\infty}^{\infty} \hat{\psi}^2_{\text{PSN,PA}}(t)\mathrm{d}t} \cdot \sqrt{\int_{-\infty}^{\infty} \psi^2_{\text{PSN,PA}}(t)\mathrm{d}t}}, \tag{7.3}$$

where $\hat{\psi}_{\text{PSN,PA}}(t)$ denotes the reconstructed channel impulse response including the LOS component. The LOS component is added to the reconstruction to make the results comparable. The scaling of the LOS component is chosen according to the energy of the measured channel impulse response. Alternatively, the scaling of the LOS component could be obtained from the distance between transmitter and receiver. However, to keep the evaluation simple, we assume the ratio of the energy of the LOS component and the energy of the remainder of the channel impulse response to be known.

Fig. 7.5 shows that the quality of the reconstructed channel impulse responses is high. In the center of the considered area, the correlation coefficient between reconstructed and measured channel impulse response is above 0.95. At the borders of the area, a correlation coefficient of about 0.8 can still be reached. However, we expect the correlation here to be

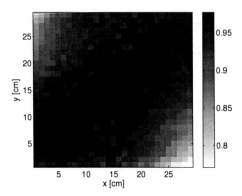

Fig. 7.5: Correlation coefficient of measured and reconstructed channel impulse responses

lower due to the limited number of training data. Furthermore, note that the reconstructed channel impulse response may contain out of band components.

To verify the practicality of the imaging based multipath prediction for UWB sensor networks, we evaluated the performance of optimized communication with the predicted multipath. We consider the generalized energy detection receiver and apply the transmitter and receiver optimization according to Chapter 3. Again we assume the ratio of the energy of the LOS component and the remainder of the channel impulse response to be given. The BER is obtained by simulation and averaged over all positions and evaluated over E_b/N_0, where E_b denotes the energy per bit and $N_0/2$ the power spectral density of the noise.

Fig. 7.6 shows the BER performance of transmitter optimization with reconstructed channel information. In this case, the receiver has a fixed post-detection filter, which is chosen as a first-order low-pass with cut-off frequency of 25 MHz. The curve marked with diamonds shows the performance of the optimization with multipath prediction. The transmit pulse is chosen according to the reconstructed channel impulse response that is generated excluding the measurement of the considered position. We use (3.10) for the optimization, where H is substituted by the Toeplitz matrix with the reconstructed channel impulse response. For comparison, the dashed line shows the transmitter optimization with full channel knowledge, the curve marked with cycles the optimization with statistical channel knowledge, and the curve with squares the case without optimization. It can be seen, that the quality of the reconstruction is high and the loss compared to full channel knowledge is small. In fact, only about 0.5 dB increase in transmit power is necessary to compensate the missing

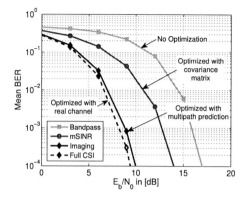

Fig. 7.6: Performance of transmitter optimization with reconstructed channel information

channel knowledge. On the other hand, the multipath prediction shows improvement over statistical channel knowledge. The optimization with statistical channel knowledge is obtained by mSINR, i.e. based on the covariance matrix and mean of the channel according to (3.19). The covariance matrix and mean are estimated based on all $M_h = 841$ channel impulse responses. The optimized pulse, which is the same for all positions, leads to an average performance loss of almost 4 dB compared to the reconstructed channel. The reason for the poor performance of the optimization with statistical channel knowledge is the structure of this sparse anechoic chamber channel. In this case, the optimization with statistical channel knowledge is not well suited and the optimization with multipath prediction performs much better. However, the optimization with statistical channel knowledge is still about 3 dB better than without optimization, where an ideal bandpass pulse is sent.

Fig. 7.7 shows the average BER performance for receiver optimization with reconstructed channel knowledge. The different cases for the optimized generalized energy detector show only marginal deviations. The dashed line with squares corresponds to the optimization of the post-detection filter with full channel knowledge. It is obtained according to (3.12) with the real channel impulse response, i.e. the maximization the SINR with full channel knowledge. The same optimization is used for the line marked with diamonds, except that the reconstructed channel impulse response is used instead of the real one. Small improvement can be achieved when using the reconstructed channel impulse response in conjunction with the optimization for statistical channel knowledge, i.e. the curve marked with cycles. In this case, the post-detection filter is obtained by (3.21), where we used the reconstructed channel

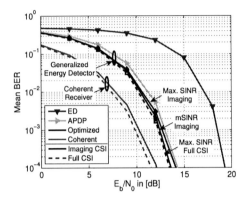

Fig. 7.7: Performance of receiver optimization with reconstructed channel information

impulse response as diagonal covariance matrix.

For comparison, the APDP receiver and the conventional energy detector (ED) with uniform integration window with 50 ns length are plotted. The ADPD is obtained by taking all $M_h = 841$ channel impulse responses into account. In addition to the non-coherent receivers, the two curves without markers show the performance for the case of a coherent detector. The dashed line uses the real channel impulse response and the solid line the reconstructed channel. The use of the reconstructed channel comes along with a loss of about 0.6 dB on average. Note that we apply an ideal bandpass filter on the receive signal to suppress the influence of out of band components in the reconstructed channel impulse response.

Overall, we observe in the simulations that the optimization with the predicted multipath performs only slightly worse than the case with full channel knowledge. The optimization shows always improvement compared to the case without optimization, where large gains can be obtained for receiver as well as transmitter optimization.

7.6 Conclusions

In this chapter, we have proposed to predict multipath delays with UWB imaging. Based on measured training data an image of the environment is generated. The image describes position and intensity of strong scatterers and reflectors. With this information, the strong multipath components can be predicted for arbitrary transmitter and receiver positions. Performance evaluation based on a controlled anechoic chamber environment prove the feasibility of the presented approach. Strong multipath can be predicted within nanoseconds accuracy. Considering UWB localization, synergies with short-range imaging can be used to facilitate high definition positioning.

Chapter 8

Measurements and Experimental Results

This chapter presents measurements and experimental results to evaluate the performance of UWB communication, localization, and imaging in a unified framework. With the previous measurement systems and data, we were so far only able to evaluate isolated parts of the system. The channel measurements that we used in the previous chapters are not perfectly suited for the comprehensive combined evaluation of localization, communication and imaging in arbitrary environments. Therefore, we discuss two advanced measurement systems in this chapter: i) a high precision measurement system and ii) an integrated low-cost/high-speed measurement system. Both systems are designed to meet the requirements for the experimental verification and performance evaluation of the algorithms presented in this thesis. Based on an extensive measurement campaign, we evaluate the accuracy of radar imaging based multipath prediction in an office environment. Furthermore, we present a selection of performance comparisons for localization and communication for UWB sensor networks in this environment. This proves the feasibility and practicality of the presented algorithms in typical indoor propagation environments.

8.1 Introduction

So far, we studied communication and localization in UWB sensor networks and introduced a family of optimized transmission schemes and timing estimation algorithms. The proposed approaches use the synergy of localization and communication and exploit the location-aware channel information. The system design is inspired by the characteristics of the wireless channel for large bandwidths. The presented results lead to insights and show the limits of low complexity UWB communication. However, for a practical implementation it is very important to compare the performance of the system to conventional approaches in a real world scenario. The benchmark should quantify the expected performance gains in typical use-cases and clearly identify the advantage of the system. A fair comparison is necessary, where all important figures of merit are covered. The best way for a practical performance assessment would be to implement the system and to try it out. However, this approach is very expensive and inflexible. For every parameter a new prototype would have to be built, which leads to high development costs.

Therefore, we evaluate the performance of the proposed schemes with computer simulation. This is a well-established approach and standard models are available. The digital signal processing can be done efficiently and analog circuits can be simulated with sufficient accuracy. This enables us to characterize the behavior of transmitter and receiver. However, special attention needs to be paid to the wireless channel. First, it is the basis of the system and the component that cannot be changed. On the other hand, it is very difficult to simulate, since it depends on the environment and on every object that is in the surrounding area.

A standard way to circumvent this problem are statistical channel models [105,106]. Some of the models are motivated by theory and others are derived from measurements. For narrowband communication, the common way of modeling the wireless channel is to scale the signal with a normally distributed random variable. This is motivated by the superposition of a large number of multipath components. With the assumption that the path gains are i.i.d. random variables, the central limit theorem leads to the normal distribution. This narrowband channel model can, however, not be used for UWB systems. With UWB, the multipath components are resolvable and do not add up. Instead, the delays of the multipath components have to be taken into account. Several methods have been proposed to characterize the distribution of the channel impulse response, see e.g. [107].

A common statistical UWB channel model is the Saleh-Valenzuela model [108], which assumes that the multipath components arrive in clusters. With the channel impulse response written as

$$h(t) = \sum_{l=0}^{L} \sum_{k=0}^{K} \alpha_{k,l} \delta(t - T_l - \tau_{k,l}),$$

the cluster arrival times are given by T_l and the relative delays of the corresponding multipath by $\tau_{k,l}$. The reasoning behind this model is that a large interacting object leads to a cluster with many single components from complex shapes in the vicinity. Moreover, the Saleh-Valenzuela model assumes that both cluster and relative delays are Poisson processes with two different arrival rates. The path gains $\alpha_{k,l}$ are normally distributed with variances decaying exponentially with cluster and relative delay. Well-known implementations of the

Saleh-Valenzuela model are the standardized IEEE 802.15.3a and IEEE 802.15.4a channel models, which are defined in [109] and [110], respectively. The first one focuses on personal area networks and covers indoor environments with LOS and NLOS situations for up to 10 m distance. The IEEE 802.15.4a channel model on the other hand includes further environments such as residential, office, outdoor, industrial, and body area networks, with each LOS and NLOS situations. The parameters of the models are obtained from extensive measurement campaigns. Realizations of channel impulse responses can be drawn according to the distributions defined in the standards.

The standardized statistical channel models are well suited to benchmark conventional UWB communication systems. Up to a certain extent, they can also be used to evaluate position estimation schemes. For instance, the randomly drawn channel impulse responses are commonly taken into account for the testing of ToA estimation algorithms. However, in our research we exploit the relationship between location and channel characteristics, which is not covered by the statistical channel models. Standard statistical channel models are not location-aware. Therefore, these models cannot be used for the performance assessment and are not suited for the presented algorithms. We require that the channel depends on the location and that the channel model is site-specific. For instance, this includes that strong multipath components can be related to objects and that there exists a relationship between location and channel characteristic.

One way to obtain a model that takes the surrounding area into account is ray-tracing. Ray-tracing is a high-frequency approximation of the wave propagation, i.e. it is assumed that objects are much greater than the wavelength. Scattering and diffraction are often included by point scatterers. Ray-tracing uses a model of the environment and determines the delay and attenuation for every object that leads to a multipath component. A reflection, e.g. of a wall, is computed by the assumption that the angle of incidence equals the angle of reflection. This way, the path lengths can be determined. Depending on the considered complexity, the ray-tracing can be restricted to single reflections, or additional higher order reflections can be taken into account. The used result of the ray-tracing is a set of multipath delays and gains that depend on the runtime of the rays and reflection coefficients of the considered objects. An implementation of ray-tracing for 2-D environments is presented e.g. in [101]. The advantage of the raytracing model is the easy study of the geometrical relationships between transmitter, receiver and surrounding objects. However, the ray-tracing model simplifies the propagation effects to a large extent and it is difficult to capture complex 3-D environments. Specifically, the main shortcomings of ray-tracing are:

• Frequency dependency of real materials is not taken into account.

- Indoor environments usually contain too many objects with too complex shapes to be captured in the model.

- The high frequency approximation does not hold for real surroundings.

- Near field and antenna effects are often not considered.

Therefore, the accuracy of ray-tracing is limited and the generated channel impulse responses may not coincide with the real channel impulse responses. In particular, most ray-tracing results underestimate the number of significant multipath components [106].

Due to the above reasons, the most suitable method to precisely predict the performance of joint UWB communication, localization and imaging algorithms is the computer simulation of transmitter and receiver based on measured channel impulse responses. So far, we used the channel measurement data from [74] and [39]. In this chapter, we present measurements and experimental results that complement the analysis of the presented algorithms. In Section 8.2, we discuss channel measurements in general and present two channel measurement systems with different objectives in Section 8.3 and 8.4. In Section 8.5, we describe an office environment measurement campaign and present selected performance results for imaging, localization and communication in Section 8.6, 8.7, and 8.8, respectively.

8.2 Channel Measurements

The goal of the channel measurement system is to obtain an estimate of the channel impulse response $h(t)$. Assuming a stationary environment, the input/output relation for the channel measurement can be modeled by

$$r(t) = h(t) * s(t) + w(t),$$

where $r(t)$ denotes the signal that can be observed at the receive antenna output. The receive signal is perturbed by white Gaussian noise $w(t)$. The input $s(t)$ denotes the signal sent by the transmit antenna. Note that we incorporate the effect of antennas, cables and other analog components into the channel.

The design of a channel measurement system is concerned with two key issues. First, it is desired to choose the transmit signal that achieves the best estimation quality. The transmit signal has a strong influence on the measurement accuracy. However, depending on the hardware of the transmitter and on the allowed transmit power, the design of the signal is subject to constraints. The second key aspect of a channel measurement system is the algorithm that is used to obtain the channel estimate $\hat{h}(t)$ from the receive signal.

Here, the choice of the estimator and the transmit signal is motivated by the following objective:

- We require the channel estimator to be unbiased, i.e.

$$\mathsf{E}\left[\hat{h}(t)\right] = h(t). \tag{8.1}$$

- The signal-to-noise ratio of the estimate should be maximized.

- Due to complexity reasons, we restrict the estimator to be a linear filter, i.e. the estimated channel response is given by

$$\hat{h}(t) = g(t) * r(t)., \tag{8.2}$$

where $g(t)$ denotes the impulse response of the linear filter and $r(t)$ the measured signal.

From (8.1) and (8.2) follows $\mathsf{E}\left[\hat{h}(t)\right] = s(t) * g(t) * h(t) \overset{!}{=} h(t)$, which is fulfilled for $s(t) * g(t) = \delta(t)$. The maximum signal-to-noise ratio is achieved if $g(t)$ is chosen as the matched filter, i.e. $g(t) = s(-t)$. Combining these two conditions results in the design criterion of the transmit signal

$$s(t) * s(-t) = \delta(t) \qquad \text{or} \qquad |S(f)|^2 = 1, \tag{8.3}$$

i.e. it is desired to transmit a signal with low correlation and accordingly white amplitude spectrum. In general, the condition (8.3) suggests to measure the channel by using a dirac delta transmit signal. However, this is not practical due to limits on the dynamic range of analog components and peak power constraints. Hence, many approximations of (8.3) have been proposed. Typical approaches can be categorized to frequency or time domain. Although both approaches lead to similar results, there are several implementation aspects to consider. Frequency domain approaches transmit chirp signals and are usually implemented by a vector network analyzer (VNA). In this case, the transmit antenna sends a frequency sweep and the amplitude and phase at the receive antenna is measured. The advantage of frequency domain approaches are the robustness against receiver non-linearities and the narrowband implementation. Since the measurement process requires a longer time, frequency domain measurements are not suited for changing environments.

In time domain, the condition (8.3) can be approximated using pseudo noise (PN) sequences. PN sequences are obtained by samples of i.i.d. random variables or they can be constructed with a structure [111–113]. Well-known examples for constructed sequences are the families of maximum length sequences (m-sequences), Gold sequences or Barker codes. The sequences are usually pulse amplitude modulated to obtain the continuous transmit signal. The base pulse should uniformly cover the bandwidth of interest without introducing inter-symbol interference.

The channel estimate is obtained from the matched filter output of the received signal, which is equivalent to the correlation with the transmit signal. The longer the signal, the better the noise attenuation. If the length of the transmit sequence is doubled, the power of the noise in the channel estimate is divided by two. This processing gain determines the gain of the signal-to-noise ratio depending on the length of the transmit sequence. The processing gain in decibel is given by $10 \log_{10} N$ if a PN sequence with N symbols is transmitted. However, very long transmit sequences require large memory and high processing power to compute the correlation. Furthermore, the length of the sequence should not extend the channel coherence time, i.e. the time the channel is considered to be not varying. The environment needs to stay constant during the channel measurement, since the channel impulse response may not change during the transmission and reception of the signals.

In the course of this thesis, we developed and used two different channel measurement systems. One system is designed to obtain channel measurements with the highest accuracy. It is built up of advanced measurement devices supporting high sensitivity to obtain reliable and reproducible measurement results. The other channel measurement system features high-speed measurements and is based on low-cost hardware. It uses standard components and performs digital signal processing on a field programmable gate array (FPGA). The system is designed to conduct channel measurement campaigns in harsh environments. The advantage is the fast processing speed, which allows to capture numerous channel impulse responses per second. This enables tracking of objects, scatterers or reflections in a moving environment such as body area networks.

8.3 High Precision Measurement System

The high precision measurement system uses an arbitrary waveform generator (AWG) at the transmitter and a digital sampling oscilloscope (DSO) at the receiver. A block diagram of the system is shown in Fig. 8.1. The detailed hardware specifications of the high precision

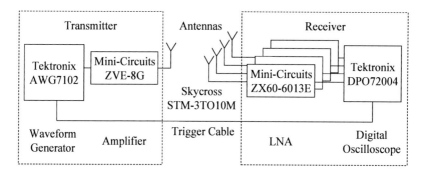

Fig. 8.1: Block diagram of High Precision Channel Measurement System

measurement system are given in Table 8.1. All components are connected by coaxial cables with SMA connectors.

The AWG generates the transmit signal from a PN sequence. The pseudo-random values are drawn independently from a normal distribution with zero mean. The sequence has a length of $2^{15} - 1$ symbols and is modulated with a symbol rate of 10 Gbps. Thus, the transmit signal has a total length of 3.27 μs. It is generated periodically and fed to the transmit amplifier. The amplifier features a high dynamic range and low noise with 30 dB gain in the band from 2 to 8 GHz. The transmit antenna is mounted on a tripod and exhibits an almost omni-directional gain pattern in the horizontal plane. The return loss is greater than 10 dB in the considered band. The same antennas are used at the receiver.

The receiver has four identical analog signal chains, which each consist of an antenna and low-noise amplifier (LNA). The LNA amplifies the receive signal up to 6 GHz by about 14 dB and compromises the accuracy of measurement minimally though its low noise figure. The LNA outputs are sampled simultaneously by the DSO with 50 GS/s each. To enable delay measurements, the AWG and DSO are synchronized via a trigger cable. The AWG sends a short pulse at the beginning of the PN sequence over this cable, which is connected to the auxiliary trigger input at the DSO. When the DSO detects the trigger, it samples the signals for 3.27 μs and saves the results.

The channel estimate is obtained from the correlation[1] of the receive signal with the template signal. The template signal is the transmit signal measured without amplifiers and antennas, i.e. the AWG output and DSO input are connected directly by cable. The auto-

---

[1]The correlation operation is computed in frequency domain using Matlab.

Device	Parameters	Values
Waveform Generator Tektronix AWG7102	Sampling frequency Resolution Clock accuracy	20 GS/s 8 bit within ±1 ppm
Transmit Amplifier Mini-Circuits ZVE-8G	Frequency range Gain Noise figure Third order intercept point (IP3)	2 − 8 GHz 30 dB ±2.0 dB 4 dB typ. +40 dBm typ.
Antenna Skycross SMT-3TO10M-A	Frequency range Horizontal pattern Size	3.1 − 10 GHz Omni-directional 16 × 13.6 × 3 mm
RF Coaxial Cables Huber+Suhner Sucoflex 104	Frequency range Length Attenuation	0 − 18 GHz 3 m, 7 m 0.5 dB/m typ.
Low Noise Amplifier Mini-Circuits ZX60-6013E	Frequency range Gain Noise figure IP3	20 MHz − 6 GHz 14 dB ±2.0 dB 3.3 dB typ. 28.7 dBm typ.
Digital Oscilloscope Tektronix DPO72004	Bandwidth Channels Sampling Rate Resolution	20 GHz 4 50 GS/s 8 bit

Table 8.1: Hardware specifications of High Precision Channel Measurement System

correlation function of the template signal is shown in Fig. 8.2 (bottom). An example of a channel impulse response is shown in Fig. 8.2 (top). It is measured in an office environment with moderate multipath. The channel impulse response includes the amplifiers and antennas. The lower bound of the frequency range is given by the transmit amplifier and antenna at about 2 GHz. At the receiver, the LNAs limit the frequency band to 6 GHz. Thus, the measurement range of the channel impulse responses is about 2 to 6 GHz.

The high precision channel measurement system provides reliable and reproducible channel measurements with a high signal-to-noise ratio. The hardware allows to measure four channel impulse responses simultaneously and synchronous to the transmitter. Thus, the channel measurement systems facilitates the evaluation of position estimation algorithms. The data acquisition, processing and storage takes less than 5 seconds for all four channels. This enables extensive measurements campaigns in a stationary environment. Further enhancements of this measurement system, including software that provides a graphical user interface, are presented in [114].

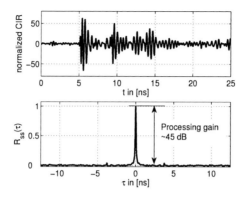

Fig. 8.2: Example of measured channel impulse response (top) and autocorrelation function of template signal (bottom).

8.4 Integrated low-cost/high-speed Measurement System

To reduce costs and duration of the channel measurements, we developed an integrated channel measurement system in a collaboration during the course of this thesis. It enables us to measure several channel impulse responses per second and can thus be used in (slowly) time-varying environments, e.g. to capture movements of humans. In particular, it facilitates the evaluation of algorithms for body area networks, such as human motion tracking [47] or radar imaging based human movement analysis [95].

Both transmitter and receiver are based on FPGAs. The advantage the FPGA-based design is its flexibility and, compared to the high-precision measurement system, the moderate cost. State-of-the-art chips offer sufficient computational power to enable fast measurement cycles and to obtain high processing gains. A block diagram of the transmitter is shown in Fig. 8.3 and of the receiver in Fig. 8.5.

The transmitter was developed by the author of [115], where further details of the design are presented. The transmitter module consists of a baseband printed circuit board (PCB), an analog radio frequency (RF) front-end PCB, and a battery. The hardware specifications of the analog RF front-end are shown in Table 8.2. The whole hardware of the transmitter node is coated in a water-resistant package to enable channel measurements in harsh environments. The battery allows autonomous operation for up to 10 hours.

The core component of the baseband PCB is the FPGA chip, where we use an Altera Cy-

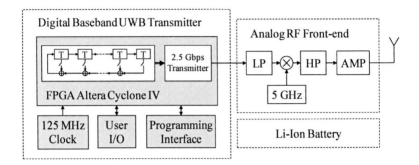

Fig. 8.3: Block diagram of integrated UWB transmitter node

Device	Parameters	Values
Low-pass Filter (LP)	Attenuation:	
	- DC − 3000 MHz	1.2 dB max.
	- 4550 MHz	20 dB min.
	- 4780 − 7500 MHz	30 dB min.
Frequency Mixer	LO/RF frequency	2.3 − 8 GHz
	IF bandwidth	DC − 3 GHz
	Conversion loss	6 dB
High Pass Filter (HP)	Attenuation:	
	- DC − 2000 MHz	20 dB min.
	- 2570 MHz	3 dB
	- 3000 − 7000 MHz	1.3 dB max.
Amplifier (AMP)	First stage:	
	- Frequency range	DC − 8 GHz
	- Gain	14 dB
	- Noise figure	4.6 dB
	- IP3	27 dBm
	Second stage:	
	- Frequency range	DC − 10 GHz
	- Gain	14 dB
	- Noise figure	7 dB
	- IP3	30 dBm

Table 8.2: Hardware specifications of Analog RF Front-end of UWB transmitter node

clone IV. This low-cost device features a 2.5 Gbps transmitter and is operated with a 125 MHz clock. Furthermore, the PCB provides a keypad, control LEDs, and a programming interface for the FPGA. The FPGA periodically generates an m-sequence, which can efficiently be

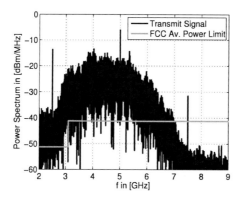

Fig. 8.4: Power spectrum of integrated UWB transmitter node

implemented by a shift register. Different sequences can be programmed and selected by the keypad. The transmitter outputs the binary pulse amplitude modulated sequence and generates a signal in the frequency range from DC to 2.5 GHz. This signal is fed to the analog front-end, where it is low-pass filtered with 3 GHz. The reason for this filter is to protect the FPGA from carrier leakage from the mixer. The signal is up-converted to 5 GHz to obtain the RF frequency range of 2.5 to 7.5 GHz. A drawback of this design is the leakage of the local oscillator frequency at 5 GHz. The carrier leakage leads to a strong peak in the spectrum of the transmitter. The DC leakage of the mixer is rejected by a high-pass filter (HP) with 2.5 GHz cutoff frequency. Subsequently, the signal is amplified in two stages of 14 dB each. The output power of the transmitter nodes (without antenna) was measured and amounts to approximately 13.7 dBm, including the carrier leakage component. The power spectrum of the transmit signal is shown in Fig. 8.4.

The integrated measurement system uses planar triangular monopole antennas, which were developed and manufactured as presented in [116]. The compact antennas have a size of $28.79\,\text{mm} \times 35.44\,\text{mm} \times 1\,\text{mm}$ and are printed on a FR-4 PCB substrate. The design is based on [117] and has been optimized towards an omni-directional pattern from 2.5 to 7.5 GHz. The antennas have a small return loss from 3 to 10.5 dB, i.e. $|S_{11}|^2 < -10$ dB. The antenna pattern was measured in an anechoic chamber and shows an approximately constant gain in the azimuth plane.

The architecture of the low-cost/high-speed receiver is shown is Fig. 8.5. The receiver consists of several analog RF front-ends and the FPGA-based digital processing. The FPGA

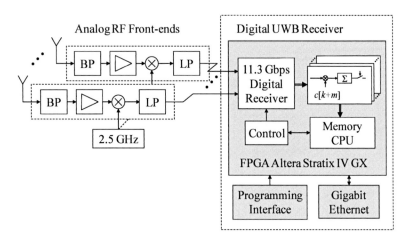

Fig. 8.5: Block diagram of integrated UWB receiver

features up to 24 receiver channels with 11.3 Gbps. We use these receiver channels as 1-bit ADC to sample the output of the analog front-ends. The details of the front-end components are specified in Table 8.3. The signal from the antenna is first fed to a band-pass filter with pass-band from 3 to 7 GHz. Then, an LNA amplifies the signal by 42 dB and a mixer down-converts it to an intermediate frequency of 2.5 GHz. Thus, the desired signal components are in the band from DC to 5 GHz. Finally, the image frequencies are suppressed by a low-pass filter.

The digital processing on the FPGA is divided into four main modules. The receiver unit acquires the signal and stores it for the massive parallel processing of the correlator units. A control unit coordinates the processing on the FPGA and an embedded CPU is responsible for the data transfer to the control PC. The details of the implementation are described in [118]. The correlator unit computes the channel estimates in time domain according to

$$\hat{h}[k] = \sum_{m=1}^{M} c[k+m]r[m], \qquad k = 1,\ldots,M,$$

where M is number of samples for one period of the transmit sequence. The template signal $c[k+m]$ is stored in the memory of the FPGA. It is constructed by the transmit sequence and up-converted to the intermediate frequency of 2.5 GHz. In the standard configuration, we use an m-sequence with $2^{15}-1$ bits. However, the receiver can easily be reconfigured

Device	Parameters	Values
Bandpass Filter (BP) Mini-Circuits VHP-26	Attenuation: - DC − 2000 MHz - 2570 MHz - 3000 − 7000 MHz	20 dB min. 3 dB 1.3 dB max.
Low-Noise-Amplifier Miteq AFS5	Frequency range Gain Noise figure 1 dB compression point	0.1 − 8 GHz 42 dB ±1.5 dB 1.4 dB max. 10 dBm
Frequency Mixer M/A-Com MY87C	LO/RF frequency IF frequency Conversion loss	0.5 − 19 GHz 0.03 − 5 GHz 7.5 dB
Low-pass Filter (LP) Mini-Circuits VLF-3400+	Attenuation: - DC − 3400 MHz - 3950 MHz - 4300 − 8300 MHz	1.5 dB max. 3 dB 20 dB min.

Table 8.3: Hardware specifications of Analog RF Front-end of integrated UWB receiver

for the use with other sequences. Due to the non-integer multiple of the FPGA sampling frequency of 11.3 GS/s and the symbol rate of 2.5 Gbps, the template signal $c[k + m]$ needs to be available for $k = 1, \ldots, M$ and for $m = 1, \ldots, M$, i.e. over two periods of the sequence. Cyclic extension of the receive signal or of the template signal would lead to wrong sampling instances. Moreover, the cyclic extension requires highly accurate symbol clocks at the transmitter and receiver. Otherwise, the cyclic extension leads to timing errors that distort the channel estimate.

The FPGA features up to 240 parallel correlators, which process up to 240 samples in a single clock cycle. With a clock frequency of 50 MHz and a sequence length of $2^{15} - 1$ bits, the computation of the full correlation process can be done in less than 8 ms. The result of the correlation are M integer values with 32 bits. The relevant part of the channel impulse response is detected by a threshold detector and the surrounding samples are transferred to the control PC. The FPGA is connected via Gigabit Ethernet to the control PC and the data transfer is coordinated by the embedded CPU. With Matlab processing on the control PC, the current setup features the acquisition and display of about 10 channel impulse responses per second. A picture of the hardware is shown in Fig. 8.6.

A limitation of the presented receiver is the 1-bit digital input. However, simulations and measurements prove that accurate and unbiased channel estimates can be obtained. Fig. 8.7 shows the linear range of the receiver and an example of an impulse response measurement. The upper figure plots the magnitude of the correlation peak, when the transmit signal was

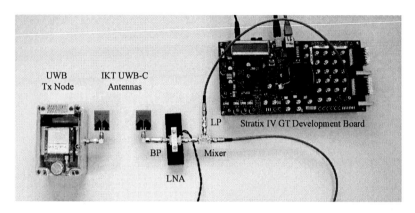

Fig. 8.6: Picture of Integrated low-cost/high-speed Measurement System

fed though a variable attenuator by cable to the receiver. It can be seen that reliable measurements are obtained over a range of at least $40\,\mathrm{dB}$, which is in line with the expected processing gain of $45\,\mathrm{dB}$ for a sequence length of $2^{15} - 1$. The lower plot depicts a measured channel impulse response together with the real channel impulse response generated by an AWG. The real and the measured channel impulse response coincide, where the measurement is slightly distorted by additive noise.

In the low-SNR regime, the clipping of the receive signal leads to an SNR increase of about $2\,\mathrm{dB}$. Due to the 1-bit ADC, the receiver is not suited for the high-SNR regime. It is important that the received signal power is lower than the noise of the LNA. If the signal exceeds the noise level, the clipping leads to irreversible errors in the channel estimate. Therefore, the low-cost/high-precision channel measurement system is not suited for high-SNR receive signals. Moreover, in the high-SNR regime the use of m-sequences is unfavorable. A multipath channel and the non-linear detector lead to additional peaks in the correlation of the sequence, see [112, 119, 120].

8.5 Office Environment Measurement Campaign

To evaluate and compare the algorithms presented in this thesis, we performed an extensive measurement campaign in an office environment. For this measurement campaign we used the high precision measurement system as it is presented in Section 8.3. The goal of the channel measurement is to obtain a large number of channel impulse responses for the sensor

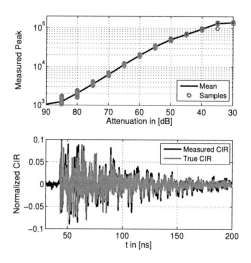

Fig. 8.7: Linearity of FPGA-based receiver and example of impulse response

network scenario. We use fixed antennas for the stationary CHs (anchors) and a mobile antenna for the SNs (agents). Besides the channel impulse responses, it is important to know the exact positions of the nodes.

As measurement scenario we chose a typical office environment. A picture of the setting is shown in Fig. 8.8. All channel measurements took place in this room[2] with the antennas mounted in a height of 1.85 m above the floor. The room is approximately rectangular and furnished with several desks, shelves and cabinets. We used the four receive antennas as the stationary anchors and the transmit antenna as agent. The agent was manually moved to $N_{SN} = 400$ different positions and the corresponding channel response to the four anchors has been measured. The total number of measured channel responses is therefore $M_h = 1600$. The channel responses have an estimated signal-to-noise-ratio of more than 45 dB.

To keep the evaluation of the measurement results as easy as possible, we chose only a 2-D setup with all positions of agents and anchors in the horizontal plane. The antennas were mounted vertically such that they radiate omni-directional in the horizontal plane. This also has the advantage that the antenna pattern can be neglected. Furthermore, the 2-D geometry simplifies the position estimation problems, since only two space variables need

---

[2]WCG laboratory, Room ETF F110

Fig. 8.8: Photo of laboratory and measurement equipment

to be considered. This way, we keep the complexity of the experimental results as small as possible. However, we still assume that we capture all effects that can be expected in the 3-D extension. Thus, we do not limit the generality and the results can be extended to 3-D in a straight-forward way.

For the assessment of position estimation and imaging algorithms it is of importance to know the real positions of the agent and anchor nodes. These can be measured manually with a distance meter. However, this procedure has several disadvantages. First, with standard equipment for manual distance measurements (measuring tape, laser rangefinder) the accuracy is limited to the range of millimeters. Second, the manual measurements are very time consuming. Third, the position of the node should be taken at the phase center of the antenna. The phase center is point of the antenna where the wave spreads spherically outwards, i.e. it is the position we should expect from unbiased position estimation. However, with a manual distance measurement it is difficult to find the phase center of the antenna and it is not clear from where to where the distance should be measured. The manual distance measurements can only be taken for the coarse estimation of the real positions.

Therefore, we decided to obtain the real positions of the agents and anchors from the ToA measurements. The channel measurements are in the high signal-to-noise-ratio regime and we used four anchors and a synchronous agent. However, without reference measurement it is not possible to evaluate the accuracy of the positions that we consider as the real positions. Still we assume the error to be insignificant, since the influence of the noise is small and with four synchronous anchors the redundancy is high.

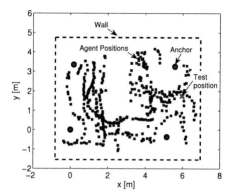

Fig. 8.9: Measurement positions and coarse floor plan of environment

To achieve the highest accuracy, the positions of the anchors and of the agent are estimated jointly. This is possible due to the synchronous measurements from $N_{SN} = 400$ agent positions. Let $\mathbf{p}_A^{(1)}, \ldots, \mathbf{p}_A^{(4)}$ denote the position of the $N_A = 4$ anchors and $T_0^{(1)}, \ldots, T_0^{(4)}$ denote the time offsets that are induced by the delays of the cables and measurement devices. Due to the synchronous measurement, the time offsets $T_0^{(1)}, \ldots, T_0^{(4)}$ are constant and do not change for different agent positions. Thus, the ToA of the LOS component holds (see Section 5.1)

$$\varepsilon_i^{(k)} = \frac{1}{c_0} \left\| \mathbf{p}_{SN}^{(i)} - \mathbf{p}_A^{(k)} \right\| + T_0^{(k)}, \qquad k = 1, \ldots, 4, \qquad i = 1, \ldots, N_{SN}.$$

The agent positions are denoted by $\mathbf{p}_{SN}^{(i)}$, where the superscript i refers to one of the N_{SN} agent positions. Assuming that the ToA estimates $\hat{\varepsilon}_i^{(k)}$ are i.i.d. Gaussian distributed, the joint ML estimator is given by

$$\hat{\boldsymbol{\theta}} = \underset{\boldsymbol{\theta} \in \mathbb{R}^{3N_A + 2N_{SN}}}{\arg\min} \sum_{i=1}^{N_{SN}} \sum_{k=1}^{N_A} \left(\hat{\varepsilon}_i^{(k)} \cdot c_0 - \left\| \mathbf{p}_{SN}^{(i)} - \mathbf{p}_A^{(k)} \right\| - c_0 T_0^{(k)} \right)^2, \tag{8.4}$$

where $\boldsymbol{\theta}$ is the set of variables

$$\boldsymbol{\theta} = \left\{ \mathbf{p}_A^{(1)}, \ldots, \mathbf{p}_A^{(4)}, T_0^{(1)}, \ldots, T_0^{(4)}, \mathbf{p}_{SN}^{(1)}, \ldots, \mathbf{p}_{SN}^{(N_{SN})} \right\}.$$

The result of the optimization is plotted in Fig. 8.9. The agent positions are marked by a cross and the anchor positions by a circle. To solve the optimization problem (8.4) we applied an iterative algorithm, which is presented in [121]. Other optimization techniques for localization with anchor position uncertainty are presented e.g. in [122–124]. For the joint estimation we used the ToA estimates obtained from WRELAX. As initial values for the anchors positions, we used the coarse measurements obtained manually with a laser rangefinder. The jointly optimized agent and anchor positions are assumed to be the real positions of the nodes. As an indicator for the accuracy of the measurements, we consider the remainder of the optimization, i.e. the objective function of the optimization problem (8.4). Since we consider only 2-D localization including time offset and have four anchor measurements available, the remainder is related to the measurement error. For every agent, we estimate the position and subtract the distances to the anchors from the pseudo-ranges and clock offsets. The RMS of these values amounts to 7.4 mm, which we assume to be the average accuracy of the presented measurements. Note that the actual measurement accuracy can be even higher, since some effects lead to a remainder in the optimization but still the right position estimate is obtained. An example for this is a deviation of the propagation speed or misadjustment of the height of the antennas.

Concluding, this measurement campaign offers a database with $M_h = 1600$ channel impulse responses including the position of transmitter and receiver and timing offset. The node positions are distributed in a room in a typical indoor office environment. This allows to simulate the performance of location-aware communication, position estimation and imaging for UWB sensor networks.

8.6 Multipath Delay Prediction

The office environment measurement campaign enables us to verify the feasibility of radar imaging based multipath delay prediction in a real world scenario. In Chapter 7, we presented the evaluation of the imaging based channel prediction based on anechoic chamber measurements. In the office environment, the reflectors are not as strong and well defined as in the anechoic chamber with metallic reflectors. Fig. 8.9 shows an approximate floor plan of the laboratory where the measurements took place. Additionally, the anchor and agent positions are plotted. In Fig. 8.10, the magnitude of the scattering coefficient map according to (7.1) is plotted. However, the walls of the room can be recognized from the image. The image is sampled with a spacing of $\Delta x = 2$ cm and $\Delta y = 1.6$ cm. In time domain, the sampling frequency has been chosen to $f_s = 50$ GHz.

As an example, Fig. 8.11 shows the measured (top) as well as reconstructed (bottom) channel response for one agent position, which is marked as test position in Fig. 8.9. The

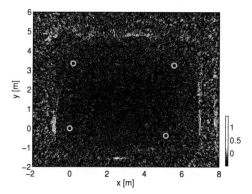

Fig. 8.10: Laboratory environment: Magnitude of scattering coefficient map

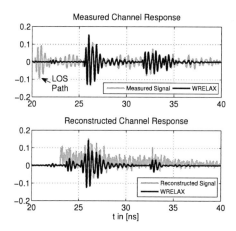

Fig. 8.11: Measured and reconstructed channel response

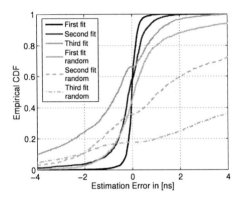

Fig. 8.12: CDF of path delay estimation error

measured channel response was excluded from training data for the reconstruction. Note that the reconstructed channel response contains out-of-band noise, which does not perturb the multipath delay estimation. Hence, we plot also the five strongest paths of both channel responses, which are extracted by WRELAX (without LOS path). It can be observed that the multipath delays of strong paths can be predicted with high accuracy.

To evaluate the accuracy of multipath delay prediction more generally, for all 400 agent positions and four different anchors the channel responses have been reconstructed. Each reconstruction was based on the training set excluding the measurements of the considered agent position. Subsequently, the five strongest multipath components have been estimated by WRELAX for the synthesized channel responses as well as for the measurement. Fig. 8.12 shows the empirical CDF of the prediction error of the multipath delay. Both predicted and measured channel response multipath extraction results in five delays. The pair with the smallest absolute error is denoted as first fit. The evaluation shows that in almost all cases at least one out of the five delays can very accurately be predicted with less than 1 ns error. The root mean square error (RMSE) is 0.298 ns. The second fit corresponds to the pair with the smallest absolute error among the remaining four delays. Here, we observe in 85.6 % an absolute error below 1 ns and accordingly for the third fit still in 47.7 % of the agents positions. For comparison, we plot the first, second and third fit of five randomly generated multipath delays, which are uniformly distributed in the interval $[t_{\text{ToA}}, t_{\text{ToA}} + 39.48\,\text{ns}]$. The time t_{ToA} denotes the time-of-arrival of the LOS path and 39.48 ns corresponds to the maximal detected excess delay of the measured channel responses. Here, the RMSE of the first

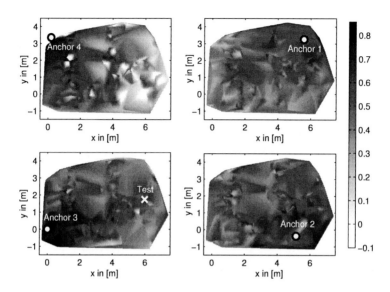

Fig. 8.13: Map of correlation coefficients between reconstructed and real channel impulse responses

fit evaluates to 2.093 ns, which is almost an order of magnitude worse than the proposed algorithm.

In Fig. 8.13, we plot the correlation coefficient between the reconstructed and real channel impulse response as a function of the position for all four anchors. For the evaluation of the correlation, we added the LOS component to the reconstruction and filtered the signal with a bandpass. The correlation coefficient is obtained according to (7.3) and results to values between −0.1 and 0.84. Generally, we observe that the reconstruction result leads to good results for agent position close to the anchor. For more distant positions, the reconstruction performs well in some cases and leads to wrong results in others. In particular, some bad results are obtained for Anchor 4. We assume that the reconstruction could be improved with a larger training set, higher positioning accuracy, and extension to the third dimension. However, for some positions, we observe that the proposed multipath prediction provides useful information of the channel. To quantify the gains of the imaging-based multipath prediction for a UWB sensor network, we evaluate the localization and communication with the predicted multipath in the following section.

8.7 Position Estimation Accuracy

Conventional range-based position estimation algorithms suffer from multipath propagation, because the LOS component cannot be distinguished from multipath. The radar-imaging based multipath prediction provides additional information about the channel for arbitrary positions in the environment. Several ways are possible to use and incorporate the additional channel knowledge into the position estimation problem. In the usual two-step approach (see Chapter 5), the location-aware channel knowledge can be included in the ToA estimation, followed by the position estimation based on the pseudo-ranges. Iteration of the two steps can then be used to refine the result.

However, to keep the performance results as general as possible, we propose a one-step approach to use the additional channel knowledge. This is based on the ML position estimation directly from the receive signals with the given channel impulse response. For the estimation, we assume the channel impulse response to be perfectly known, i.e. we do not account for reconstruction errors. To evaluate the position estimation accuracy, we consider two different cases: i) remote-localization, and ii) self-localization. For remote-localization, the position of the SN is estimated at the CHs. Due to the less stringent complexity constraints of the CHs, we assume that the position estimation can be based on the perfectly sampled receive signals. On the other hand, the self-localization considers the position estimation at the sensor nodes. Therefore, self-localization is based on the output of a generalized energy detection receiver.

First, we consider the case of *remote-localization*. Specifically, we assume that the receive signal at anchor $i = 1, \ldots, N_A$ from sensor position p_{SN} is given by

$$r^{(i)}(t) = h_{\mathsf{p}_{SN}}^{(i)}(t - T_0) * s(t) + n_i(t), \qquad (8.5)$$

with i.i.d. white Gaussian noise $n_i(t)$. Based on these receive signals, the ML position estimation problem can be written as

$$\left\{ \hat{\mathsf{p}}_{SN}, \hat{T}_0 \right\} = \operatorname*{arg\,max}_{\mathsf{p}_{SN} \in \mathbb{R}^2, T_0 \in \mathbb{R}} \; p\left(r^{(1)}(t), \ldots, r^{(N_A)}(t) | \mathsf{p}_{SN}, T_0 \right). \qquad (8.6)$$

With i.i.d. Gaussian noise, the solution to (8.6) is given by

$$\left\{ \hat{\mathsf{p}}_{SN}, \hat{T}_0 \right\} = \operatorname*{arg\,min}_{\mathsf{p}_{SN} \in \mathbb{R}^2, T_0 \in \mathbb{R}} \; \sum_{i=1}^{N_A} \int_{-\infty}^{\infty} \left(r^{(i)}(t) - h_{\mathsf{p}_{SN}}^{(i)}(t - T_0) * s(t) \right)^2 \mathrm{d}t.$$

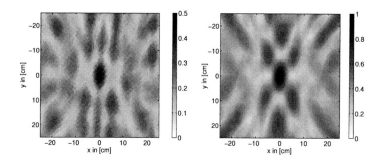

Fig. 8.14: Maximum sum correlation for reconstructed channel knowledge (left) and full channel knowledge (right) for different positions.

Expanding the squares leads to

$$\left\{\hat{\mathbf{p}}_{\mathrm{SN}}, \hat{T}_0\right\} = \underset{\mathbf{p}_{\mathrm{SN}}\in\mathbb{R}^2, T_0\in\mathbb{R}}{\arg\max} \sum_{i=1}^{N_A} \int_{-\infty}^{\infty} r^{(i)}(t)\left[h_{\mathrm{PSN}}^{(i)}(t-T_0)*s(t)\right]\mathrm{d}t,$$

where we assume that $\int_{-\infty}^{\infty}|h_{\mathrm{PSN}}^{(i)}(t)|^2\mathrm{d}t$ is independent of $\mathbf{p}_{\mathrm{SN}}$. In frequency domain, the optimization problem can equivalently be written as

$$\left\{\hat{\mathbf{p}}_{\mathrm{SN}}, \hat{T}_0\right\} = \underset{\mathbf{p}_{\mathrm{SN}}\in\mathbb{R}^2, T_0\in\mathbb{R}}{\arg\max} \sum_{i=1}^{N_A} \int_{-\infty}^{\infty} R^{(i)}(f)H_{\mathrm{PSN}}^{(i)}(f)S(f)e^{j2\pi fT_0}\mathrm{d}f, \tag{8.7}$$

where $R^{(i)}(f)$, $H_{\mathrm{PSN}}^{(i)}(f)$, $S(f)$ are the Fourier transform of $r^{(i)}(t)$, $h_{\mathrm{PSN}}^{(i)}(t)$, and $s(t)$, respectively.

The ML estimator corresponds to the maximum search over the sum of the correlation for the expected receive signal at different positions. Note that the optimization problem (8.7) is difficult to solve, because it usually has many maximums. This can be seen in Fig. 8.14, where we plot

$$\max_{T_0}\sum_{i=1}^{N_A} \int_{-\infty}^{\infty} R^{(i)}(f)H_{\mathrm{PSN}}^{(i)}(f)S(f)e^{j2\pi fT_0}\mathrm{d}f,$$

without noise and the signals normalized to unit energy as a function of the deviation from the sensor node position $\mathbf{p}_{\mathrm{SN}}$. The receive signals $R^{(i)}(f)$ are the Fourier transform of the measured channel impulse at the test position marked in Fig. 8.9. In the left plot, we substitute $H_{\text{PSN}}^{(i)}(f)$ with the reconstructed channel impulse responses, which are complemented with the LOS component, bandpass filtered and normalized to unit energy. It can be seen, that the maximum correlation coefficient is obtained for the true position with a value of about 0.5. For comparison, Fig. 8.9 (right) shows the maximum correlation coefficient for a case with full channel knowledge. Here, we assume that the shape of the channel impulse response is constant and only the delay depends on the position, i.e. we substitute the channel impulse response with

$$H_{\text{PSN}}^{(i)}(f) = H^{(i)}(f) \cdot e^{-j2\pi f \frac{1}{c_0} \left\| \mathbf{P_{SN}} - \mathbf{P_A}^{(i)} \right\|},$$

where $H^{(i)}(f)$ are the measured channel impulse responses at the test position. This case shows the performance bound of the position estimation accuracy and is not practical due to the missing location-dependency of the channel. We observe a maximum correlation of 1 at the true position and many sidelobes with smaller magnitude. The reason for the sidelobes are the high correlation of multipath components with the LOS component at a certain delay.

For *self-localization*, the position estimation is based on the sampled output of the generalized energy detector, i.e.

$$y_i(lT_s) = \int\limits_{-\infty}^{\infty} g(\tau)\tilde{r}_i^2(lT_s - \tau)\mathrm{d}\tau, \qquad\qquad l = 1, \ldots, L$$

where $\tilde{r}_i^2(lT_s - \tau)$ is the bandpass filtered and squared receive signal (8.5). To keep the problem tractable, we assume $y(lT_s)$ to be independent and normally distributed. The position can then be estimated according to

$$\left\{\hat{\mathbf{p}}_{\text{SN}}, \hat{T}_0\right\} = \underset{\mathbf{p_{SN}} \in \mathbb{R}^2, T_0 \in \mathbb{R}}{\arg\max} \sum_{i=1}^{N_A} \sum_{l=1}^{L} y_i(lT_s) \int\limits_{-\infty}^{\infty} g(\tau) \left(\tilde{h}_{\text{PSN}}^{(i)}(t - T_0) * s(t) \Big|_{t=lT_s - \tau} \right)^2 \mathrm{d}\tau \tag{8.8}$$

For the case of imaging based channel knowledge, the channel impulse responses $\tilde{h}_{\text{PSN}}^{(i)}(t)$ are substituted with the normalized and bandpass filtered reconstructions including the LOS component. With full channel knowledge, we consider again the time-shifted channel impulse responses according to the distance to the anchor, which does not take the location-dependent shape into account.

To assess and compare the accuracy of position estimation with imaging information, we evaluated (8.7) and (8.8) with noisy receive signals. For the simulation, we add white

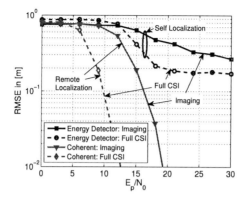

Fig. 8.15: Comparison of positioning error with real and reconstructed channel knowledge for coherent receiver and generalized energy detector

Gaussian noise with power spectral density $N_0/2$ to the receive pulse with energy E_p. As figure-of-merit, we evaluate the RMS of the position estimation error for varying signal-to-noise ratios E_p/N_0. The result is shown in Fig. 8.15, where we compare remote-localization and self-localization for imaging as well as full channel information. We always consider the test position, which is marked in Fig. 8.9. To solve the optimization problems (8.7) and (8.8), we apply a grid search. Due to the high complexity of the evaluation of the cost function, only a limited number of grid points can be considered. For the search over the timing offset T_0 we use sampling with a frequency of 50 GHz, which corresponds to a spatial resolution of about 0.6 cm. The search over space is conducted on a uniform grid of 2×2 m around the test position with 2 cm resolution. This leads to 10000 position hypotheses, where the maximum sum correlation is evaluated for each position hypothesis. The maximizing grid point is considered as the solution to the estimation problem.

The performance of remote-localization, i.e. with a coherent receiver with full rate sampling, is marked with triangles and diamonds. The dashed line shows the RMSE with full channel information and the solid line with the reconstructed channels. We observe about 6 dB penalty for the reconstructed channels, which is in line with the maximum sum correlation coefficient of about 0.5 as shown in Fig. 8.14. Note that the real position is included in the grid and that for high SNR always the real position is estimated, which leads to a vanishing RMSE. Due to the discrete grid it is not possible to resolve positioning errors that are smaller than the grid resolution. The grid resolution of 2 cm corresponds to a remaining RMSE of 0.81 cm.

The curves marked with squares and cycles show the performance of sensor self-localization. In this case, we consider a generalized energy detection receiver with first-order low-pass filter with cutoff frequency of $300\,\text{MHz}$. The output is sampled with $1/T_s = 1\,\text{GHz}$ leading to an observation window of $L = 160$ samples. Whereas we obtain a remaining RMSE of about $17.1\,\text{cm}$ for full channel information at high SNR, the performance with reconstructed channels saturates at about $26.5\,\text{cm}$. This shows that with a non-coherent receiver the penalty of imperfect channel information is small. Additional anchors could be used to further improve the positioning accuracy enabling high definition positioning for sensor self-localization. The simulation results show that imaging-based reconstructed channel knowledge enables localization even with a very low complexity receiver without high rate sampling.

8.8 Performance of Location-aware Communication

To further demonstrate the benefit of imaging-based multipath prediction, we evaluate the performance of location-aware communication with predicted channel information in the office environment. In Section 7.5, we have already shown that promising performance gains can be obtained in the anechoic chamber with two metallic reflectors. In this section, we extend the performance analysis to a more typical environment and show the feasibility of location-aware communication with imaging-based multipath prediction.

As it can be seen in Fig. 8.13, the multipath prediction does not produce proper results for all positions. In some cases even negative correlation coefficients between the reconstruction and the measured channel impulse response are obtained, which would lead to a severe performance degradation for transmitter as well as receiver optimization. Due to the dominance of these single outage events, the BER performance averaged over all positions would suffer strongly and hide the gains of the multipath prediction. However, to still being able to verify the potential of imaging-based multipath prediction, we restrict the analysis to a single link, i.e. the link between the test position and Anchor 3, see Fig. 8.13. This channel shows strong multipath components that can successfully be predicted from the imaging data. We applied the transmitter and receiver optimization according the Chapter 3 for this test position and evaluated the BER by simulation.

In Fig. 8.16, the performance of transmitter optimization is plotted in terms of BER versus E_b/N_0. The post-detection filter of the generalized energy detection receiver is again chosen as first-order low-pass filter with cutoff frequency of $25\,\text{MHz}$. For comparison, we

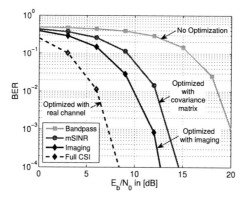

Fig. 8.16: Example for transmitter optimization with imaging-based channel knowledge

consider four different cases: No optimization (marked with squares), optimized with the covariance matrix (marked with cycles), optimized with imaging (solid line marked with diamonds), and optimized with the real channel (dashed line). The optimized transmit pulse with the real channel is obtained according to (3.10), i.e. maximum SINR with full channel knowledge. The same optimization method is used for the case with imaging, where the real channel impulse response is substituted with the bandpass filtered predicted channel including the LOS component. We observe a penalty of about 4.4 dB due to the imperfect channel state information. However, the optimization with the imaging-based predicted multipath shows still about 2 dB gain compared to the optimization based on the covariance matrix and mean of the channel. In the latter case, we optimize the transmit pulse according to maximization of the mean SINR (mSINR), i.e. the pulse is obtained with (3.19). Finally, without optimization an ideal bandpass pulse is transmitted, which does not require any channel state information. This leads to a performance loss of more than 8 dB compared to location-aware communication with imaging-based predicted multipath.

Fig. 8.17 shows the BER versus E_b/N_0 for receiver optimization. The results for the generalized energy detection receiver are plotted with markers and for comparison for a coherent receiver without markers. In all cases, the transmitter uses an ideal bandpass pulse. For the generalized energy detection receiver, the solid line with cycles and diamonds show the performance of the receiver optimization with the imaging-based predicted multipath. The curve with diamonds corresponds to the maximization of the SINR according to (3.12), where we substituted the predicted channel impulse response. For the curve with cycles, we

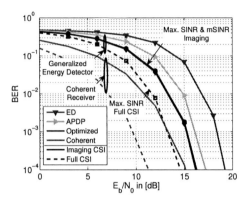

Fig. 8.17: Example for receiver optimization with imaging-based channel knowledge

used the maximization of mean SINR (mSINR), i.e. (3.21) with $\Sigma_h = \mathrm{diag}(\tilde{h}_{\mathrm{PSN}} \odot \tilde{h}_{\mathrm{PSN}})$, where $\tilde{h}_{\mathrm{PSN}}$ is the vector with the predicted channel impulse response. Both approaches show marginal performance differences and about 1 dB gain compared to the APDP receiver. The APDP receiver uses the post-detection filter with the power delay profile at anchor 3 averaged over all measurement positions. Compared to the conventional energy detector (ED) with uniform integration window, the imaging-based approaches show more than 3 dB gain. On the other hand, the dashed line with squares shows the BER of the receiver optimization with the real channel, which is about 1.5 dB better than the imaging-based approaches. For the coherent receiver, the performance loss with the predicted channel is larger and amounts to approximately 4 dB. This corresponds to a correlation coefficient of 0.65 and is in line with the value plotted in Fig. 8.13.

8.9 Summary

In this chapter, we present measurements and experimental results to quantify the gains of joint communication, localization and imaging for UWB sensor networks. We present two channel measurement systems that were developed and used during this thesis. The first measurement system is designed for high-precision measurements with up to four anchors. Additionally, we present an FPGA-based system, to enable short measurement cycles with a high update rate. With the high-precision system, an extensive channel measurement campaign has be conducted in an office environment. The channel data enables us to evaluate the imaging-based multipath prediction and shows that channel impulse responses can successfully be reconstructed from the training data. Furthermore, we verify and demonstrate the practicality of position estimation with predicted multipath and location-aware communication. We show that the additional channel information can be used to improve the system performance and thus save channel estimation overhead.

Chapter 9

Conclusions and Outlook

9.1 Conclusions

UWB technology is a promising candidate for the physical layer of future wireless sensor networks. The large bandwidth enables reliable short-range communication in harsh propagation environments as well as localization and imaging. In particular, non-coherent UWB communication is the method-of-choice for low complexity, low power and low cost systems. A smart implementation of non-coherent UWB is the generalized energy detection receiver. Its advantages are the robustness to channel variations and the low complexity analog implementation, which does not require high rate sampling of the receive signal. In this thesis, we contribute to communication, localization, and imaging for UWB sensor networks with generalized energy detection receivers. In summary, we draw the following conclusions.

Result 1. *Communication in UWB sensor networks benefits from* a priori *CSI.*

This is the conclusion of the first part, where we develop a framework for the optimization of UWB communication in sensor networks with *a priori* channel information. We introduce location-aware communication, i.e. the transceiver is adapted to the multipath channel conditions depending on the position of the sensor node. In particular, we consider generalized energy detection receivers with pulse position modulation. Conventionally, these receivers show moderate detection performance and are vulnerable to narrowband interference. Therefore, we derive transmitter and receiver optimization schemes based on the Signal-to-Interference-plus-Noise-Ratio (SINR): First, we present the SINR optimization based on full channel knowledge. A priori CSI is then incorporated by means of a statistical channel model that depends on the position of the sensor node. Performance evaluation based on a simple channel model is used to give insight about the fundamental behavior of the derived optimization schemes. Moreover, an extensive performance evaluation with channel data from a rich scattering environment proves that location information can improve the data transmission and helps to successfully suppress narrowband interference. Performance gains of 2 to 5 dB compared to conventional energy detection can be obtained.

In the next step, we consider multiuser precoding for the downlink of UWB sensor networks, i.e. communication from a central unit to many sensor nodes. In harsh propagation environments, the detection performance strongly suffers from inter-symbol interference due to multipath, which substantially limits the data rate or requires expensive receiver postprocessing. To overcome this problem, we propose a novel precoding scheme to transmit to several nodes simultaneously. This way the sum data rate can be increased, while low complexity of sensor nodes is maintained. First, precoding optimization is derived for this setup based on full channel state information, where we maximize the minimum signal quality over all nodes. In a second step, optimization is extended to statistical channel knowledge, which depends on the position of the nodes. Performance evaluation shows that multiple nodes can efficiently be served simultaneously. Only marginal increase in transmit power is necessary compared to time-multiplexing.

Result 2. *UWB position estimation benefits from* a priori *CSI.*

This conclusion is drawn in the second part of this thesis, which studies localization in UWB sensor networks. In particular, we consider range-based position estimation and focus on the timing estimation of the receive signal. Standard timing estimation algorithms are not applicable for sensor self-localization due to stringent requirements on complexity and power consumption of the receiver. Therefore, we derive the maximum likelihood timing estimation for the generalized energy detection receiver with a priori channel knowledge. We show that the estimation accuracy can be improved by using the exact marginal PDF of the energy detector output under the assumption that the channel impulse response is a Gaussian random process. This requires only the mean and the covariance matrix of the channel to be known at the receiver. Furthermore, we systematically derive low-complexity timing estimation algorithms from approximations of the exact PDF. Performance evaluation based on a channel model and measurements quantifies the accuracy of the proposed algorithms and shows that ranging with centimeter precision can be achieved at the low complexity sensor nodes.

Additionally, we propose spectral timing estimation at the generalized energy detector output, which can be implemented with lowest complexity and power consumption. The advantage of this approach is that it uses no high-speed ADC and requires only low sampling rates for the digital processing. Besides the basic idea to apply spectral timing estimation at the energy detector output, the main contribution is the analysis of this estimator under multipath propagation. We show that this system benefits from resolvability of multipath components due to a large signaling bandwidth and draw a comparison to narrowband phase estimation. Performance analysis of the proposed estimator yields an accuracy of up to 20 cm in strong multipath environments.

Result 3. *Use UWB radar image to generate vicinity based* a priori *CSI.*

In the third part of this thesis, we introduce radar imaging based multipath prediction as a method to generate the *a priori* CSI. This establishes an efficient way to obtain the relationship between the position and the channel characteristics. A set a training data with measured channel impulse response is sufficient generate a map of the surrounding propagation environment. The knowledge of the reflecting and scattering objects is used to predict the channel impulse response for arbitrary positions. We propose a low-complexity imaging algorithms and show that impulse responses can be predicted with high accuracy. We demonstrate and verify the radar imaging based multipath delay prediction in a defined anechoic chamber environment and prove that the vicinity based *a priori* CSI can successfully improve location-aware UWB communication.

To demonstrate radar imaging multipath prediction in typical sensor network environments, we discuss two advanced channel measurement systems. Both are designed to facilitate the verification and benchmark of joint UWB communication, localization and imaging algorithms, whereas one aims for high-precision and the other for low-cost/high-speed measurements. Finally, an extensive measurement campaign in an office environment enables us to quantify the gains of synergetic communication and localization in a UWB sensor networks. Selected performance results are presented to prove the feasibility of imaging-based multipath prediction and its use for localization and location-aware communication.

9.2 Outlook on Future Research

In this thesis, we provide a framework for communication and localization in UWB sensor networks. We deliver a family of algorithms to optimize data transmission, to improve short range position estimation, to sense the environment and thus provide channel knowledge by multipath prediction. This toolbox aims to provide insights and guidelines for the designer of future UWB sensor networks. However, further research is necessary before this technology becomes ready for the market. Additionally, we see great potential for further research and innovations in the fascinating field of UWB communication, localization, and imaging. Therefore, in the following we give a brief overview on open topics, which the author considers as the most relevant and interesting.

Integrated circuit design: While signal processing and the specifics of UWB communication are well-investigated in theory, it is also important to develop suitable hardware in order to support the large bandwidth of UWB technology. Today's RF microelectronics design focuses on mixed-signal circuits for conventional narrowband systems. Additional research is necessary to improve the analysis, design and optimization of integrated UWB technology. The large bandwidth makes high demands on analog components such as filters, amplifiers and mixers. In particular, the digital processing with very high sampling rates requires high complexity and power consumption with standard semiconductor technologies and devices. To circumvent this problem, we propose to use the generalized energy detection receiver. With further research on the chip design for UWB, we expect promising implementations of this receiver that can meet the stringent requirements of wireless sensor networks.

Localization: We expect a high demand and profitable future market for high definition localization with UWB technology. UWB localization systems can be implemented with low costs and are a key-enabler for a variety of leading-edge innovations and smart products. The unique feature of multipath robustness makes UWB localization to one of the most promising applications of UWB technology. Besides the implementation of the well-known algorithms, we expect a high potential for further research in the area of UWB localization. In particular, we consider the design and analysis of cooperative localization schemes as relevant and interesting. Today, most position estimation approaches are restricted to hierarchic networks with anchors and agents. However, many open issues arise for relative positioning in large networks without infrastructure. Distributed localization in systems without anchors is a research topic that can disruptively improve the practicability and accuracy of UWB localization. Future research should optimize strategies and algorithms for cooperative localization maximizing the availability and accuracy while minimizing the complexity and required number of transmissions or range measurements.

Communication: Today's research on UWB sensor networks aims to achieve coverage with low complexity, low power and low costs sensor nodes. In the future, we expect a large increase in node density and data traffic. The growing number of nodes requires further means to manage the interfering data transmissions that will limit the network performance. Therefore, further research on UWB sensor networks should consider multiuser communication as well as cooperative strategies. In particular, the combination of powerful transmission schemes with stringent requirements on complexity and power consumption bears the highest potential for seminal innovations.

Imaging: A crucial part of the sensing of the propagation environment is the applied imaging algorithms. The exact solution of the wave equations requires tremendous complexity and is impractical for the implementation. Therefore, it is important to find good trade-offs between complexity and performance. We believe that the radar imaging based multipath prediction can be improved by more accurate imaging algorithms, which are too complex for today's implementation. A promising approach to circumvent this problem could be object recognition or to include further a priori knowledge to simplify the prediction of multipath. Additionally, motion models should be taken into account to track the changes of the environment.

Besides the fundamental research, we believe that application-specific innovations and products will lead to the breakthrough of UWB technology. During the course of this thesis, we considered different applications, which can strongly benefit from UWB technology. In the following, we give a list of some applications that are very promising in the author's view.

Body Area Networks: UWB technology enables human motion tracking with high accuracy and low costs. UWB transceivers included in clothes can be used to analyze the movement and the posture of the body, see e.g. [47]. Furthermore, radar-imaging based tracking of the multipath delays facilitates fall detection for elderly or athletes, see [95].

Security System: UWB distance bounding is a technology to enable secure localization. This is of major importance for access control and authentication. Conventional narrow-band systems are vulnerable to relaying attacks [125]. The short pulses of UWB enable RTT measurements with very short delays to establish a tight upper bound on the distance measurement, see [48].

Industrial sensor networks: UWB communication facilitates sensor data transmission in harsh propagation environments with large metallic objects or sealed areas. Therefore, UWB is perfectly suited for machine and engine telemetry and can be used e.g. in ship's engines to collect data from temperature or pressure sensors. The advantage is the easy installation and maintenance free operation. In combination with localization the sensors can be replaced without reprogramming.

Machine tools: Non-coherent UWB communication has great potential for the wireless data transmission in machine tools. Conventional, it is very difficult to transfer sensor data from fast moving parts in CNC machines such as drills or lathes. This can be solved by non-coherent UWB impulse radio due to its robustness to channel variations, small size and low power consumption.

A Derivation of (6.14)-(6.16)

Expression (6.14) *and* (6.16): Let $\gamma \sim \mathcal{U}[0,T]$ and $BT \in \mathbb{N}$, then

$$\mathsf{E}\left[\sin\left(2\pi\frac{\gamma}{T}\right)^2\right] = \mathsf{E}\left[\sin\left(\pi\gamma B\right)^2\right] = \int_0^T \frac{1}{T}\sin\left(2\pi\frac{\gamma}{T}\right)^2 \mathrm{d}\gamma$$

$$= \frac{1}{T}\left[\frac{\gamma}{2} - \frac{T\sin\left(4\pi\frac{\gamma}{T}\right)}{8\pi}\right]_0^T = \frac{1}{2}$$

Expression (6.15): Let $\gamma_n \sim \mathcal{U}[0,T]$, $\gamma_m \sim \mathcal{U}[0,T]$ i.i.d. and $BT \in \mathbb{N}$, then

$$I_{n,m} = \mathsf{E}\left[\left(\frac{\sin\left(\pi(\gamma_n+\gamma_m)\frac{1}{T}\right)\sin\left(\pi(\gamma_n-\gamma_m)B\right)}{\sin\left(\pi(\gamma_n-\gamma_m)\frac{1}{T}\right)}\right)^2\right]$$

$$= \int_{-\infty}^{\infty}\int_{-\infty}^{\infty} p_{\gamma_n,\gamma_m}(\gamma_n,\gamma_m)\left(\frac{\sin\left(\pi(\gamma_n+\gamma_m)\frac{1}{T}\right)\sin\left(\pi(\gamma_n-\gamma_m)B\right)}{\sin\left(\pi(\gamma_n-\gamma_m)\frac{1}{T}\right)}\right)^2 \mathrm{d}\gamma_n\mathrm{d}\gamma_m$$

$$= \frac{1}{T^2}\int_0^T\int_0^T\left(\frac{\sin\left(\pi(\gamma_n+\gamma_m)\frac{1}{T}\right)\sin\left(\pi(\gamma_n-\gamma_m)B\right)}{\sin\left(\pi(\gamma_n-\gamma_m)\frac{1}{T}\right)}\right)^2 \mathrm{d}\gamma_n\mathrm{d}\gamma_m,$$

where we used $p_{\gamma_n,\gamma_m}(\gamma_n,\gamma_m) = p_{\gamma_n}(\gamma_n)p_{\gamma_m}(\gamma_m) = \frac{1}{T^2}$ for $0 \leq \gamma_n \leq T$ and $0 \leq \gamma_m \leq T$. With the substitution

$$x = \frac{\pi}{T}(\gamma_n - \gamma_m) \qquad \Leftrightarrow \qquad \gamma_n = \frac{T}{2\pi}(x+y)$$
$$y = \frac{\pi}{T}(\gamma_n + \gamma_m) \qquad\qquad\qquad \gamma_m = \frac{T}{2\pi}(y-x)$$

and

$$|J| = \left|\frac{\partial\gamma_n}{\partial x}\frac{\partial\gamma_m}{\partial y} - \frac{\partial\gamma_n}{\partial y}\frac{\partial\gamma_m}{\partial x}\right| = \frac{T^2}{2\pi^2},$$

we find

$$
I_{n,m} = \frac{1}{2\pi^2} \int\limits_{y=0}^{\pi} \int\limits_{x=-y}^{y} \frac{\sin^2(y)\sin^2(BTx)}{\sin^2(x)} \, \mathrm{d}x\mathrm{d}y
$$

$$
+ \frac{1}{2\pi^2} \int\limits_{y=\pi}^{2\pi} \int\limits_{x=-2\pi+y}^{2\pi-y} \frac{\sin^2(y)\sin^2(BTx)}{\sin^2(x)} \, \mathrm{d}x\mathrm{d}y
$$

$$
= \frac{2}{\pi^2} \int\limits_{y=0}^{\pi} \int\limits_{x=0}^{y} \frac{\sin^2(y)\sin^2(BTx)}{\sin^2(x)} \, \mathrm{d}x\mathrm{d}y.
$$

The last line follows from periodicity of the integrand. The integral over x computes to

$$
\int\limits_{x=0}^{y} \frac{\sin^2(BTx)}{\sin^2(x)} \, \mathrm{d}x = \sum_{n=1}^{BT} y + \frac{BT-n}{n} \sin(2ny), \qquad \text{for } BT \in \mathbb{N}
$$

This can be shown as follows:

$$
\frac{\partial}{\partial y} \left(\sum_{n=1}^{K} y + \frac{K-n}{n} \sin(2ny) \right)
$$

$$
= \sum_{n=1}^{K} 1 + 2(K-n)\cos(2ny)
$$

$$
= \frac{1-\cos(2y)}{2\sin^2(y)} \sum_{n=1}^{K} 1 + 2(K-n)\cos(2ny)
$$

$$
= \frac{1}{2\sin^2(y)} \sum_{n=1}^{K} 1 - \cos(2y) + 2(K-n)\cos(2ny) - 2(K-n)\cos(2ny)\cos(2y)
$$

$$
= \frac{1}{2\sin^2(y)} \left(K - K\cos(2y) + \sum_{n=1}^{K} 2(K-n)\cos(2ny) \right.
$$

$$
\left. - \sum_{n=1}^{K} (K-n)\cos(2(n-1)y) - \sum_{n=1}^{K} (K-n)\cos(2(n+1)y) \right)
$$

$$
= \frac{1}{2\sin^2(y)} \left(K - K\cos(2y) + \sum_{n=1}^{K} 2(K-n)\cos(2ny) \right.
$$

$$
\left. - \sum_{n=0}^{K-1} (K-n-1)\cos(2ny) - \sum_{n=2}^{K+1} (K-n+1)\cos(2ny) \right)
$$

$$= \frac{1}{2\sin^2(y)} \left(K - K\cos(2y) + \sum_{n=1}^{K} 2(K-n)\cos(2ny) \right.$$

$$- (K-1) - \cos(2Ky) - \sum_{n=1}^{K} (K-n-1)\cos(2ny)$$

$$\left. +K\cos(2y) - \sum_{n=1}^{K} (K-n+1)\cos(2ny) \right)$$

$$= \frac{1}{2\sin^2(y)} \left(1 - \cos(2Ky) + \sum_{n=1}^{K} \underbrace{[2(K-n) - (K-n-1) - (K-n+1)]}_{=0}\cos(2ny) \right)$$

$$= \frac{1 - \cos(2Ky)}{2\sin^2(y)}$$

$$= \frac{\sin^2(Ky)}{\sin^2(y)}$$

Finally, we have

$$I_{n,m} = \frac{2}{\pi^2} \int_{y=0}^{\pi} \sin^2(y) \left(\sum_{n=1}^{BT} y + \frac{BT-n}{n} \sin(2ny) \right) \mathrm{d}y$$

$$= \frac{2BT}{\pi^2} \int_{y=0}^{\pi} y\sin^2(y)\mathrm{d}y + \frac{2}{\pi^2} \sum_{n=1}^{BT} \frac{BT-n}{n} \underbrace{\int_{y=0}^{\pi} \sin^2(y)\sin(2ny)\mathrm{d}y}_{=0}$$

$$= \frac{2BT}{\pi^2} \left[\frac{y^2}{4} - \frac{1}{8}\cos(2y) - \frac{1}{4}y\sin(2y) \right]_{y=0}^{\pi}$$

$$= \frac{2BT}{\pi^2} \frac{\pi^2}{4} = \frac{BT}{2}.$$

Acronyms

ADC	analog-to-digital converter
AoA	angle of arrival
APDP	average power delay profile
AWG	arbitrary waveform generator
AWGN	additive white Gaussian noise
BER	bit error rate
BPPM	binary pulse position modulation
CDF	cumulative distribution function
CH	cluster head
CIR	channel impulse response
CSI	channel state information
DGPS	differential GPS
DSO	digital sampling oscilloscope
ED	energy detector
FFT	fast Fourier transform
FPGA	field programmable gate array
GNSS	global navigation satellite system
i.i.d.	independent and identically distributed
IP3	third order intercept point
IPDP	instantaneous power delay profile
IQ	in-phase and quadrature

IR   impulse radio
ISI   inter-symbol interference

LNA   low-noise amplifier
LOS   line-of-sight
LTI   linear time invariant

m-sequence maximum length sequence
MEMS   microelectromechanical systems
ML   maximum likelihood

NLOS   non-line-of-sight

OFDM   orthogonal frequency-division multiplexing
OOK   on-off-keying

PCB   printed circuit board
PDF   probability density function
PDP   power delay profile
PN   pseudo noise

RF   radio frequency
RMS   root mean square
RMSE   root mean square error
RSS   received signal strength
RTT   round trip time
RX   receiver

SAR   synthetic aperture radar
SINR   signal-to-interference-plus-noise-ratio
SN   sensor node
SNR   signal-to-noise ratio

TDMA   time division multiple access
ToA   time of arrival
TX   transmitter

UWB   ultra-wideband

VLSI   very-large-scale integration

Notation

$\mathbf{a}$	column vector
$\mathbf{A}$	matrix
$[\mathbf{A}]_{i,j}$	(i,j)-th element of matrix $\mathbf{A}$
$[\mathbf{A}]_{n:m,k:l}$	submatrix of $\mathbf{A}$ formed by rows n to m and columns k to l
$\arg(z)$	phase angle of complex number z
c_0	speed of light in air (≈ 299702547 m/s)
$*$	convolution
$\mathrm{diag}(\mathbf{a})$	diagonal matrix with the vector $\mathbf{a}$ on its diagonal
$\delta(t)$	Dirac delta function
$\mathsf{E}[\cdot]$	expectation operator
$\nabla_{\mathbf{x}}$	gradient vector with respect to $\mathbf{x}$
$\odot$	Hadamard product
$H_{\mathbf{x}}$	Hessian matrix with respect to $\mathbf{x}$
$\mathbf{I}_N$	identity matrix of dimension N
$\mathrm{Im}\{z\}$	imaginary part of complex number z
$[a,b]$	closed interval of real numbers between a and b
j	$\sqrt{-1}$
$\otimes$	Kronecker product

$\lambda_{\max}\{\mathbf{A},\mathbf{B}\}$ generalized principle eigenvalue of matrix $\mathbf{A}$ and $\mathbf{B}$, i.e. the generalized eigenvalue of largest magnitude

$\arg\max$ maximizing argument of a function or set

$\arg\min$ minimizing argument of a function or set

$\mathbb{N}$ set of natural numbers

$\|\cdot\|_1$ ℓ_1-norm, sum of the absolute values

$\|\cdot\|$ ℓ_2-norm, Euclidean norm

$\mathcal{N}\left(\mu,\sigma^2\right)$ Normal distribution with mean μ and variance σ^2

π constant Pi (≈ 3.14159)

$\Pr[\cdot]$ probability of an event

$\Pr[\cdot|\cdot]$ conditional probability of an event

$Q(x)$ $\frac{1}{2}\mathrm{erfc}(\frac{x}{\sqrt{2}})$, where $\mathrm{erfc}(\cdot)$ denotes the complementary error function

$\mathbb{R}$ set of real numbers

$\mathrm{sinc}(x)$ $\frac{\sin(\pi x)}{\pi x}$

T_p pulse repetition period

T_{ppm} time shift of the pulse for BPPM, usually $T_{\mathrm{symb}}/2$

T_{symb} symbol duration

$(\cdot)^T$ transpose of a matrix or vector

$\mathrm{Tr}[\mathbf{A}]$ sum of diagonal elements of matrix $\mathbf{A}$

$\mathcal{U}(a,b)$ continuous uniform distribution in the interval $[a,b]$

$\mathrm{Var}[\cdot]$ variance of a random variable

$\mathbf{v}_{\max}\{\mathbf{A}\}$ principle eigenvector of matrix $\mathbf{A}$, i.e. the eigenvector corresponding to the eigenvalue of largest magnitude

$\mathbf{v}_{\max}\{\mathbf{A},\mathbf{B}\}$ generalized principle eigenvector of matrix $\mathbf{A}$ and $\mathbf{B}$, i.e. the generalized eigenvector corresponding to the generalized eigenvalue of largest magnitude

$\mathbb{Z}$ set of integers

Bibliography

[1] M. G. Di Benedetto, T. Kaiser, A. F. Molisch, I. Oppermann, C. Politano, and D. Por-cino, *UWB Communication Systems–A Comprehensive Overview*. Hindawi Publish-ing Corporation, 2006.

[2] L. Yang and G. B. Giannakis, "Ultra-wideband communications: an idea whose time has come," *IEEE Signal Processing Mag.*, vol. 21, no. 6, pp. 26–54, Nov. 2004.

[3] D. Porcino and W. Hirt, "Ultra-wideband radio technology: potential and challenges ahead," *IEEE Commun. Mag.*, vol. 41, no. 7, pp. 66–74, July 2003.

[4] G. Falciasecca, "Marconi's early experiments in wireless telegraphy, 1895," *IEEE Trans. Antennas Propag.*, vol. 52, no. 6, pp. 220–221, Dec. 2010.

[5] F. Gardiol and Y. Fournier, "Marconi in Switzerland: True story or fairy tale?" in *Proc. IEEE History of Telecommunications Conf., HISTELCON 2008,*, Sept. 2008, pp. 12–19.

[6] M. Win and R. Scholtz, "Impulse radio: how it works," *IEEE Commun. Lett.*, vol. 2, no. 2, pp. 36–38, Feb. 1998.

[7] M. Chiani and A. Giorgetti, "Coexistence between UWB and narrow-band wireless communication systems," *Proc. IEEE*, vol. 97, no. 2, pp. 231–254, Feb. 2009.

[8] M. Win and R. Scholtz, "On the robustness of ultra-wide bandwidth signals in dense multipath environments," *IEEE Commun. Lett.*, vol. 2, no. 2, pp. 51–53, Feb. 1998.

[9] ——, "Ultra-wide bandwidth time-hopping spread-spectrum impulse radio for wire-less multiple-access communications," *IEEE Trans. Commun.*, vol. 48, no. 4, pp. 679–689, Apr. 2000.

[10] N. Beaulieu and D. Young, "Designing time-hopping ultrawide bandwidth receivers for multiuser interference environments," *Proc. IEEE*, vol. 97, no. 2, pp. 255–284, Feb. 2009.

[11] FCC, "Revision of part 15 of the commission's rules regarding ultra-wideband transmission systems," *1st Report and Order, ET Docket 98-153, FCC 02-48*, adopted/released Feb. 14/Apr. 22 2002.

[12] ECMA-368, "High rate ultra wideband PHY and MAC standard," 3rd edition, Dec. 2008.

[13] Hewlett-Packard, Intel, LSI, Microsoft, NEC, and S. ST-Ericsson, "Wireless universal serial bus," Specification 1.1, Sept. 2010.

[14] IEEE 802.15.4a-2007, "Part 15.4: Wireless medium access control (MAC) and physical layer (PHY) specifications for low-rate wireless personal area networks (WPANs); amendment 1: Add alternate PHYs," Aug. 2007.

[15] J. Zhang, P. Orlik, Z. Sahinoglu, A. Molisch, and P. Kinney, "UWB systems for wireless sensor networks," *Proc. IEEE*, vol. 97, no. 2, pp. 313–331, Feb. 2009.

[16] Z. Ahmadian and L. Lampe, "Performance analysis of the IEEE 802.15.4a UWB system," *IEEE Trans. Commun.*, vol. 57, no. 5, pp. 1474–1485, May 2009.

[17] S. Gezici, Z. Tian, G. Giannakis, H. Kobayashi, A. Molisch, H. Poor, and Z. Sahinoglu, "Localization via ultra-wideband radios: a look at positioning aspects for future sensor networks," *IEEE Signal Processing Mag.*, vol. 22, no. 4, pp. 70–84, July 2005.

[18] S. Gezici and H. Poor, "Position estimation via ultra-wide-band signals," *Proc. IEEE*, vol. 97, no. 2, pp. 386–403, Feb. 2009.

[19] K. Pahlavan, X. Li, and J. Makela, "Indoor geolocation science and technology," *IEEE Commun. Mag.*, vol. 40, no. 2, pp. 112–118, Feb. 2002.

[20] J.-Y. Lee and R. Scholtz, "Ranging in a dense multipath environment using an UWB radio link," *IEEE J. Sel. Areas Commun.*, vol. 20, no. 9, pp. 1677–1683, Dec. 2002.

[21] D. Dardari, A. Conti, U. Ferner, A. Giorgetti, and M. Win, "Ranging with ultrawide bandwidth signals in multipath environments," *Proc. IEEE*, vol. 97, no. 2, pp. 404–426, Feb. 2009.

[22] W. Foy, "Position-location solutions by taylor-series estimation," *IEEE Trans. Aerosp. Electron. Syst.*, vol. AES-12, no. 2, pp. 187–194, Mar. 1976.

[23] P. Bahl and V. Padmanabhan, "RADAR: an in-building RF-based user location and tracking system," in *Proc. 19th Annu. Joint Conf. of the IEEE Computer and Communications Societies, INFOCOM 2000*, vol. 2, Mar. 2000, pp. 775–784.

[24] C. Steiner, F. Althaus, F. Trösch, and A. Wittneben, "Ultra-wideband geo-regioning: A novel clustering and localization technique," *EURASIP Journal on Advances in Signal Processing*, Nov. 2007.

[25] C. Steiner and A. Wittneben, "Low complexity location fingerprinting with generalized UWB energy detection receivers," *IEEE Trans. Signal Process.*, vol. 58, no. 3, pp. 1756–1767, Mar. 2010.

[26] L. Jofre, A. Broquetas, J. Romeu, S. Blanch, A. Toda, X. Fabregas, and A. Cardama, "UWB tomographic radar imaging of penetrable and impenetrable objects," *Proc. IEEE*, vol. 97, no. 2, pp. 451–464, Feb. 2009.

[27] E. Fear, P. Meaney, and M. Stuchly, "Microwaves for breast cancer detection?" *IEEE Potentials*, vol. 22, no. 1, pp. 12–18, Feb. 2003.

[28] R. Zetik, J. Sachs, and R. Thoma, "UWB short-range radar sensing – the architecture of a baseband, pseudo-noise UWB radar sensor," *IEEE Instrum. Meas. Mag.*, vol. 10, no. 2, pp. 39–45, Apr. 2007.

[29] J. A. Scales, *Theory of Seismic Imaging*. Springer-Verlag Berlin Heidelberg, 1995.

[30] S. Nag, M. A. Barnes, T. Payment, and G. Holladay, "Ultrawideband through-wall radar for detecting the motion of people in real time," *Proc. SPIE*, vol. 4744, no. 1, pp. 48–57, 2002.

[31] E. Kaasinen, "User needs for location-aware mobile services," *Personal and Ubiquitous Computing*, vol. 7, pp. 70–79, 2003.

[32] Y. Xu, J. Heidemann, and D. Estrin, "Geography-informed energy conservation for ad hoc routing," in *Proc. ACM MobiCom '01*, New York, NY, USA, 2001, pp. 70–84.

[33] D. Huang, M. Mehta, D. Medhi, and L. Harn, "Location-aware key management scheme for wireless sensor networks," in *Proc. ACM SASN '04*, New York, NY, USA, 2004, pp. 29–42.

[34] N. Patwari, J. Ash, S. Kyperountas, I. Hero, A.O., R. Moses, and N. Correal, "Locating the nodes: cooperative localization in wireless sensor networks," *IEEE Signal Processing Mag.*, vol. 22, no. 4, pp. 54–69, Jul. 2005.

[35] H. Wymeersch, J. Lien, and M. Win, "Cooperative localization in wireless networks," *Proc. IEEE*, vol. 97, no. 2, pp. 427–450, Feb. 2009.

[36] N. Bulusu, J. Heidemann, and D. Estrin, "GPS-less low-cost outdoor localization for very small devices," *IEEE Pers. Commun.*, vol. 7, no. 5, pp. 28–34, Oct. 2000.

[37] T. Zasowski, "A system concept for ultra wideband UWB body area networks," Ph.D. dissertation, ETH Zurich, 2007.

[38] F. Troesch, "Novel low duty cycle schemes: From ultra wide band to ultra low power," Ph.D. dissertation, ETH Zurich, 2009.

[39] C. Steiner, "Location fingerprinting for ultra-wideband systems – the key to efficient and robust localization," Ph.D. dissertation, ETH Zurich, 2010.

[40] H. Luecken, T. Zasowski, C. Steiner, F. Troesch, and A. Wittneben, "Location-aware adaptation and precoding for low complexity IR-UWB receivers," in *Proc. IEEE Int. Conf. on Ultra-Wideband, ICUWB 2008*, vol. 3, Sept. 2008, pp. 31–34.

[41] H. Luecken, T. Zasowski, and A. Wittneben, "Synchronization scheme for low duty cycle UWB impulse radio receiver," in *Proc. IEEE Int. Symp. on Wireless Communication Systems, ISWCS 2008*, Oct. 2008, pp. 503–507.

[42] H. Luecken, C. Steiner, and A. Wittneben, "ML timing estimation for generalized UWB-IR energy detection receivers," in *Proc. IEEE Int. Conf. on Ultra-Wideband, ICUWB 2009*, Sept. 2009, pp. 829–833.

[43] H. Luecken and A. Wittneben, "Low complexity positioning system for indoor multipath environments," in *Proc. IEEE Int. Conf. on Communications, ICC 2010*, May 2010.

[44] ——, "Multiuser precoding for UWB sensor networks," in *Proc. IEEE Int. Symp. on Wireless Communication Systems, ISWCS 2011*, Nov. 2011, pp. 156–160.

[45] ——, "UWB radar imaging based multipath delay prediction for NLOS position estimation," in *Proc. IEEE Int. Conf. on Ultra-Wideband, ICUWB 2011*, Sep. 2011, pp. 101–105.

[46] H. Luecken, C. Steiner, and A. Wittneben, "Location-aware UWB communication with generalized energy detection receivers," *IEEE Trans. Wireless Commun.*, vol. 11, no. 9, pp. 3068–3078, Sept. 2012.

[47] Z. Mekonnen, E. Slottke, H. Luecken, C. Steiner, and A. Wittneben, "Constrained maximum likelihood positioning for UWB based human motion tracking," in *Proc. Int. Conf. on Indoor Positioning and Indoor Navigation, IPIN 2010*, Sept. 2010.

[48] M. Kuhn, H. Luecken, and N. O. Tippenhauer, "UWB impulse radio based distance bounding," in *Proc. 7th Workshop on Positioning Navigation and Communication, WPNC 2010*, Mar. 2010, pp. 28–37.

[49] C. Steiner, H. Luecken, T. Zasowski, F. Troesch, and A. Wittneben, "Ultra low power UWB modem design: Experimental verification and performance evaluation," in *Union Radio Scientifique Internationale, URSI 2008*, Aug. 2008.

[50] I. Akyildiz, W. Su, Y. Sankarasubramaniam, and E. Cayirci, "A survey on sensor networks," *IEEE Commun. Mag.*, vol. 40, no. 8, pp. 102–114, Aug. 2002.

[51] H. Cao, V. Leung, C. Chow, and H. Chan, "Enabling technologies for wireless body area networks: A survey and outlook," *IEEE Commun. Mag.*, vol. 47, no. 12, pp. 84–93, Dec. 2009.

[52] A. Willig, K. Matheus, and A. Wolisz, "Wireless technology in industrial networks," *Proc. IEEE*, vol. 93, no. 6, pp. 1130–1151, Jun. 2005.

[53] A. Ganguly, K. Chang, S. Deb, P. Pande, B. Belzer, and C. Teuscher, "Scalable hybrid wireless network-on-chip architectures for multicore systems," *IEEE Trans. Comput.*, vol. 60, no. 10, pp. 1485–1502, Oct. 2011.

[54] I. F. Akyildiz and J. M. Jornet, "Electromagnetic wireless nanosensor networks," *Nano Communication Networks*, vol. 1, no. 1, pp. 3–19, Mar. 2010.

[55] C. Savarese, J. Rabaey, and J. Beutel, "Location in distributed ad-hoc wireless sensor networks," in *Proc. IEEE Int. Conf. on Acoustics, Speech and Signal Processing, ICASSP 2001*, vol. 4, May 2001, pp. 2037–2040.

[56] J. Choi and W. Stark, "Performance of ultra-wideband communications with suboptimal receivers in multipath channels," *IEEE J. Sel. Areas Commun.*, vol. 20, no. 9, pp. 1754–1766, Dec. 2002.

[57] M. Weisenhorn and W. Hirt, "ML receiver for pulsed UWB signals and partial channel state information," in *Proc. IEEE Int. Conf. on Ultra-Wideband, ICU 2005*, Sept. 2005, pp. 379–384.

[58] F. Troesch, C. Steiner, T. Zasowski, T. Burger, and A. Wittneben, "Hardware aware optimization of an ultra low power UWB communication system," in *Proc. IEEE Int. Conf. on Ultra-Wideband, ICUWB 2007*, Sept. 2007, pp. 174–179.

[59] X. Luo, L. Yang, and G. Giannakis, "Designing optimal pulse-shapers for ultrawideband radios," *Journal of Communications and Networks*, vol. 5, no. 4, pp. 344–353, Dec. 2003.

[60] B. Hu and N. Beaulieu, "Pulse shapes for ultrawideband communication systems," *IEEE Trans. Wireless Commun.*, vol. 4, no. 4, pp. 1789–1797, Jul. 2005.

[61] Y. Zhu and H. Wu, "Integrated distributed transversal filters for pulse shaping and interference suppression in UWB impulse radios," in *Proc. IEEE Int. Conf. on Ultra-Wideband, ICUWB 2006*, Sep. 2006, pp. 563–568.

[62] T. Strohmer, M. Emami, J. Hansen, G. Papanicolaou, and A. Paulraj, "Application of time-reversal with MMSE equalizer to UWB communications," in *Proc. IEEE Global Telecommunications Conf., Globecom 2004*, vol. 5, Nov. 2004, pp. 3123–3127.

[63] H. T. Nguyen, I. Kovacs, and P. Eggers, "A time reversal transmission approach for multiuser UWB communications," *IEEE Trans. Antennas Propag.*, vol. 54, no. 11, pp. 3216–3224, Nov. 2006.

[64] N. Guo, B. Sadler, and R. Qiu, "Reduced-complexity UWB time-reversal techniques and experimental results," *IEEE Trans. Wireless Commun.*, vol. 6, no. 12, pp. 4221–4226, Dec. 2007.

[65] Y.-H. Chang, S.-H. Tsai, X. Yu, and C.-C. Kuo, "Performance enhancement of channel-phase precoded ultra-wideband (CPP-UWB) systems by rake receivers," in *Proc. IEEE Global Telecommunications Conf., Globecom 2008*, Nov. 2008.

[66] ——, "Codeword design for ultra-wideband (UWB) precoding," *IEEE Trans. Wireless Commun.*, vol. 9, no. 1, pp. 198–207, Jan. 2010.

[67] A.-H. Mohsenian-Rad, J. Mietzner, R. Schober, and V. Wong, "Pre-equalization for pre-rake DS-UWB systems with spectral mask constraints," *IEEE Trans. Commun.*, vol. 59, no. 3, pp. 780–791, Mar. 2011.

[68] A. D'Amico, U. Mengali, and E. Arias-de Reyna, "Energy-detection UWB receivers with multiple energy measurements," *IEEE Trans. Wireless Commun.*, vol. 6, no. 7, pp. 2652–2659, July 2007.

[69] T. Zasowski, F. Troesch, and A. Wittneben, "Partial channel state information and intersymbol interference in low complexity UWB PPM detection," in *Proc. IEEE Int. Conf. on Ultra-Wideband, ICUWB 2006*, Sept. 2006, pp. 369–374.

[70] T. Zasowski and A. Wittneben, "Performance of UWB receivers with partial CSI using a simple body area network channel model," *IEEE J. Sel. Areas Commun.*, vol. 27, no. 1, pp. 17–26, Jan. 2009.

[71] L. Isserlis, "On a formula for the product-moment coefficient of any order of a normal frequency distribution in any number of variables," *Biometrika*, vol. 12, no. 1-2, pp. 134–139, 1918.

[72] K. Witrisal, G. Leus, G. Janssen, M. Pausini, F. Troesch, T. Zasowski, and J. Romme, "Noncoherent ultra-wideband systems," *IEEE Signal Processing Mag.*, vol. 26, no. 4, pp. 48–66, July 2009.

[73] S. Dubouloz, B. Denis, S. de Rivaz, and L. Ouvry, "Performance analysis of LDR UWB non-coherent receivers in multipath environments," in *Proc. IEEE Int. Conf. on Ultra-Wideband, ICU 2005*, Sept. 2005, pp. 491–496.

[74] F. Althaus, F. Troesch, T. Zasowski, and A. Wittneben, "STS measurements and characterization," *PULSERS Deliverable D3b6a*, vol. IST-2001-32710 PULSERS, 2005.

[75] L. Stoica, A. Rabbachin, H. Repo, T. Tiuraniemi, and I. Oppermann, "An ultrawideband system architecture for tag based wireless sensor networks," *IEEE Trans. Veh. Technol.*, vol. 54, no. 5, pp. 1632–1645, Sept. 2005.

[76] F. Troesch, T. Zasowski, and A. Wittneben, "Non-linear UWB receivers with MLSE post-detection," in *Proc. IEEE Vehicular Technology Conf., VTC Spring 2008*, May 2008.

[77] Y.-H. Chang, S.-H. Tsai, X. Yu, and C.-C. J. Kuo, "Ultrawideband transceiver design using channel phase precoding," *IEEE Trans. Signal Process.*, vol. 55, no. 7, pp. 3807–3822, July 2007.

[78] J. Wagner, "Distributed forwarding in multiuser multihop wireless networks," Ph.D. dissertation, ETH Zurich, 2011.

[79] C. Carbonelli and U. Mengali, "Synchronization algorithms for UWB signals," *IEEE Trans. Commun.*, vol. 54, no. 2, pp. 329–338, Feb. 2006.

[80] A. Rabbachin and I. Oppermann, "Synchronization analysis for UWB systems with a low-complexity energy collection receiver," in *Proc. Int. Workshop on Ultra Wideband Systems, UWBST & IWUWBS 2004*, May 2004, pp. 288–292.

[81] I. Guvenc, Z. Sahinoglu, and P. Orlik, "TOA estimation for IR-UWB systems with different transceiver types," *IEEE Trans. Microw. Theory Tech.*, vol. 54, no. 4, pp. 1876–1886, June 2006.

[82] S. Aedudodla, S. Vijayakumaran, and T. Wong, "Timing acquisition in ultra-wideband communication systems," *IEEE Trans. Veh. Technol.*, vol. 54, no. 5, pp. 1570–1583, Sept. 2005.

[83] J. Li and R. Wu, "An efficient algorithm for time delay estimation," *IEEE Trans. Signal Process.*, vol. 46, no. 8, pp. 2231–2235, Aug. 1998.

[84] G. Tziritas, "On the distribution of positive-definite Gaussian quadratic forms," *IEEE Trans. Inf. Theory*, vol. 33, no. 6, pp. 895–906, Nov. 1987.

[85] U. Grenander, H. O. Pollak, and D. Slepian, "The distribution of quadratic forms in normal variates: A small sample theory with applications to spectral analysis," *Journal of the Society for Industrial and Applied Mathematics*, vol. 7, no. 4, pp. 374–401, 1959.

[86] H. Meyr, M. Moeneclaey, and S. Fechtel, *Digital Communication Receivers: Synchro-nization, Channel Estimation, and Signal Processing.* New York, NY, USA: John Wiley & Sons, Inc., 1997.

[87] S. Kotz, N. Balakrishnan, and N. L. Johnson, *Continuous Multivariate Distributions, Volume 1, Models and Applications.* New York, NY, USA: John Wiley & Sons, Inc., 2000.

[88] H. Fan and C. Yan, "Asynchronous differential TDOA for sensor self-localization," in *Proc. IEEE Int. Conf. on Acoustics, Speech and Signal Processing, ICASSP 2007*, vol. 2, Apr. 2007, pp. II–1109–II–1112.

[89] E. D. Kaplan, *Understanding GPS: Principles and Applications.* Artech House, 1996.

[90] D. Dardari, C.-C. Chong, and M. Win, "Threshold-based time-of-arrival estimators in UWB dense multipath channels," *IEEE Trans. Commun.*, vol. 56, no. 8, pp. 1366–1378, Aug. 2008.

[91] M. Oerder and H. Meyr, "Digital filter and square timing recovery," *IEEE Trans. Commun.*, vol. 36, no. 5, pp. 605–612, May 1988.

[92] A. C. Rencher and G. B. Schaalje, *Linear Models in Statistics*, 2nd ed. John Wiley & Sons, Inc, 2008.

[93] R. Fontana, "Recent system applications of short-pulse ultra-wideband (UWB) tech-nology," *IEEE Trans. Microw. Theory Tech.*, vol. 52, no. 9, pp. 2087–2104, Sep. 2004.

[94] Y. Yang and A. Fathy, "See-through-wall imaging using ultra wideband short-pulse radar system," in *IEEE Antennas and Propagation Society International Symposium 2005*, vol. 3B, Jul. 2005, pp. 334–337.

[95] T. Rüegg, "UWB radar imaging based human movement analysis and posture detec-tion," Master's thesis, ETH Zurich, 2012.

[96] J. Gazdag and P. Sguazzero, "Migration of seismic data," *Proc. IEEE*, vol. 72, no. 10, pp. 1302–1315, Oct. 1984.

[97] W. A. Schneider, "Integral formulation for migration in two and three dimensions," *Geophysics*, vol. 43, no. 1, pp. 49–76, 1978.

[98] X. Zhuge, T. Savelyev, A. Yarovoy, L. Ligthart, and B. Levitas, "Comparison of dif-ferent migration techniques for UWB short-range imaging," in *Proc. European Radar Conf., EuRAD 2009*, Oct. 2009, pp. 184–187.

[99] X. Zhuge, T. Savelyev, A. Yarovoy, and L. Ligthart, "UWB array-based radar imaging using modified kirchhoff migration," in *Proc. IEEE Int. Conf. on Ultra-Wideband, ICUWB 2008*, vol. 3, Sep. 2008, pp. 175–178.

[100] X. Zhuge, A. Yarovoy, T. Savelyev, and L. Ligthart, "Modified kirchhoff migration for UWB MIMO array-based radar imaging," *IEEE Trans. Geosci. Remote Sens.*, vol. 48, no. 6, pp. 2692–2703, Jun. 2010.

[101] A. Bracher, D. Sutter, and X. Zhang, "UWB imaging system: Reconstruction algorithm and demonstrator," ETH Zurich, Group Project, 2010.

[102] Y. Shen and M. Win, "Fundamental limits of wideband localization – Part I: A general framework," *IEEE Trans. Inf. Theory*, vol. 56, no. 10, pp. 4956–4980, Oct. 2010.

[103] P. Meissner, C. Steiner, and K. Witrisal, "UWB positioning with virtual anchors and floor plan information," in *Proc. 7th Workshop on Positioning Navigation and Communication, WPNC 2010*, Mar. 2010, pp. 150–156.

[104] C. Steiner and A. Wittneben, "Efficient training phase for ultrawideband-based location fingerprinting systems," *IEEE Trans. Signal Process.*, vol. 59, no. 12, pp. 6021 –6032, dec. 2011.

[105] A. F. Molisch, "Ultrawideband propagation channels-theory, measurement, and modeling," *IEEE Trans. Veh. Technol.*, vol. 54, pp. 1528–1545, Sept. 2005.

[106] A. Molisch, "Ultra-wide-band propagation channels," *Proc. IEEE*, vol. 97, no. 2, pp. 353–371, Feb. 2009.

[107] U. G. Schuster, H. Boelcskei, and G. Durisi, "Ultra-wideband channel modeling on the basis of information-theoretic criteria," in *Proc. IEEE Int. Symp. on Information Theory, ISIT*, Sept. 2005, pp. 97–101.

[108] A. Saleh and R. Valenzuela, "A statistical model for indoor multipath propagation," *IEEE J. Sel. Areas Commun.*, vol. 5, no. 2, pp. 128–137, Feb. 1987.

[109] Jeff R. Foerster et. al., "Channel modeling sub-comitee report final," *IEEE P802.15 WG for WPANs Technical Report, no. 02/490r0-SG3a*, 2002.

[110] A. F. Molisch, K. Balakrishnan, C.-C. Chong, S. Emami, A. Fort, J. Karedal, J. Kunisch, H. Schantz, U. Schuster, and K. Siwiak, "IEEE 802.15.4a channel model - final report," IEEE P802.15 02/490r1–SG3a, Tech. Rep., Sept. 2004.

[111] M. R. Schroeder, "Integrated-impulse method measuring sound decay without using impulses," *The Journal of the Acoustical Society of America*, vol. 66, no. 2, pp. 497–500, 1979.

179

[112] D. D. Rife and J. Vanderkooy, "Transfer-function measurement with maximum-length sequences," *J. Audio Eng. Soc.*, vol. 37, no. 6, pp. 419–444, Jun. 1989.

[113] H. D. Lüke, *Korrelationssignale.* Springer, Berlin, 1992.

[114] G. Hauzenberger, "Wireless UWB sensors for marine engine telemetry," Master's thesis, ETH Zurich, 2012.

[115] C. Sulser, "UWB transmit nodes," ETH Zurich, Hardware Presentation, 2011.

[116] ——, "Antenna designs UWB / RacooN," ETH Zurich, IKT WCS-R Presentation, 2011.

[117] C.-C. Lin and H.-R. Chuang, "A 3-12 GHz UWB planar triangular monopole antenna with ridged ground-plane," *Progress In Electromagnetics Research*, vol. 83, pp. 307–321, 2008.

[118] S. Schürch, "Implementation of an FPGA based receiver for UWB channel measurements," ETH Zurich, Semester Thesis, 2012.

[119] C. Dunn and M. Hawksford, "Distortion immunity of MLS-derived impulse response measurements," *J. Audio Eng. Soc.*, vol. 41, no. 5, pp. 314–335, May 1993.

[120] M. Greest and M. Hawksford, "Nonlinear distortion analysis using maximum-length sequences," *Electronics Letters*, vol. 30, no. 13, pp. 1033–1035, Jun. 1994.

[121] Z. W. Mekonnen, "Self-calibration algorithm for TOA-based localization systems," ETH Zurich, Technical Report, 2012.

[122] K. Lui, W.-K. Ma, H. So, and F. Chan, "Semi-definite programming algorithms for sensor network node localization with uncertainties in anchor positions and/or propagation speed," *IEEE Trans. Signal Process.*, vol. 57, no. 2, pp. 752–763, Feb. 2009.

[123] ——, "Semi-definite programming approach to sensor network node localization with anchor position uncertainty," in *Proc. IEEE Int. Conf. on Acoustics, Speech and Signal Processing, ICASSP 2009*, Apr. 2009, pp. 2245–2248.

[124] G. Shirazi, M. Shenouda, and L. Lampe, "Second order cone programming for sensor network localization with anchor position uncertainty," in *Proc. 8th Workshop on Positioning Navigation and Communication, WPNC 2011*, Apr. 2011, pp. 51–55.

[125] A. Francillon, B. Danev, and S. Capkun, "Relay attacks on passive keyless entry and start systems in modern cars," in *Proc. of the Network and Distributed System Security Symp., NDSS 2011.* The Internet Society, 2011.

Curriculum Vitae

Name:	Heinrich Luecken
Birthday:	May 20, 1981
Birthplace:	Aachen, Germany

Education

10/2007-12/2012	**ETH Zurich, Switzerland** PhD studies at the Communication Technology Laboratory, Department of Information Technology and Electrical Engineering
09/2001-07/2007	**RWTH Aachen University, Germany** Electrical Engineering, Degree: Dipl.-Ing.
08/2004-06/2005	**Royal Institute of Technology (KTH), Stockholm, Sweden** Exchange student
06/2000	**Freie Waldorfschule Aachen, Germany** Abitur

Experience

10/2007-03/2013	**ETH Zurich, Switzerland** Research assistant at the Communication Technology Laboratory, Prof. Dr. Armin Wittneben
06/2006-12/2006	**Robert Bosch RTC, Palo Alto, CA, USA** Wireless Technologies Intern
10/2005-05/2006	**Chair of Electrical Engineering and Computer Systems (EECS), RWTH Aachen, Germany** Student research assistant, Prof. Dr. Tobias G. Noll
02/2003-04/2003	**Marconi Ondata, Backnang, Germany** Optical Network and Architecture Intern

Awards

11/2008	Friedrich-Wilhelm-Preis for diploma thesis "Development of quantitative models for a cost-benefit analysis of a generic satellite navigation receiver"
09/2008	Springorum-Denkmünze for Dipl.-Ing. degree with highest honors

Publications

1. **Location-aware UWB Communication with Generalized Energy Detection Receivers**
   H. Luecken, C. Steiner, and A. Wittneben, *IEEE Transactions on Wireless Communications*, Vol. 11, No. 9, pp. 3068-3078, Sept. 2012

2. **Multiuser Precoding for UWB Sensor Networks**
   H. Luecken and A. Wittneben, *IEEE International Symposium on Wireless Communication Systems*, ISWCS 2011, Aachen, Germany, Nov. 2011

3. **UWB Radar Imaging based Multipath Delay Prediction for NLOS Position Estimation**
   H. Luecken and A. Wittneben, *IEEE International Conference on Ultra-Wideband*, ICUWB 2011, Bologna, Italy, Sept. 2011

4. **Constrained Maximum Likelihood Positioning for UWB Based Human Motion Tracking**
   Z. W. Mekonnen, E. Slottke, H. Luecken, C. Steiner, and A. Wittneben, *International Conference on Indoor Positioning and Indoor Navigation*, IPIN 2010, Zurich, Switzerland, Sept. 2010

5. **Low Complexity Positioning System for Indoor Multipath Environments**
   H. Luecken and A. Wittneben, *IEEE International Conference on Communications*, ICC 2010, Cape Town, South Africa, May 2010

6. **Attacks on Physical-layer Identification**
   B. Danev, H. Luecken, S. Capkun, and K. El Defrawy, *ACM conference on Wireless network security*, WiSec 2010, Hoboken NJ, USA, Mar. 2010

7. **UWB Impulse Radio Based Distance Bounding**
   M. Kuhn, H. Luecken, and N. O. Tippenhauer, *Workshop on Positioning, Navigation and Communication*, WPNC 2010, Dresden, Germany, Mar. 2010

8. **ML Timing Estimation for Generalized UWB-IR Energy Detection Receivers**
   H. Luecken, C. Steiner, and A. Wittneben, *IEEE International Conference on Ultra-Wideband*, ICUWB 2009, Vancouver, Canada, Sept. 2009

9. **Hardware-in-the-Loop Simulator for Performance and Cost Estimation of GNSS Correlator Channels**
   G. Kappen, H. Luecken, L. Kurz, and T. G. Noll, *European Navigation Conference - Global Navigation Satellite Systems*, ENC-GNSS 2009, Naples, Italy, May 2009

10. **Synchronization Scheme for Low Duty Cycle UWB Impulse Radio Receiver**
    H. Luecken, T. Zasowski, and A. Wittneben, *IEEE International Symposium on Wireless Communication Systems*, ISWCS 2008, Reykjavik, Iceland, Oct. 2008

11. **Location-aware Adaptation and Precoding for Low Complexity IR-UWB Receivers**
    H. Luecken, T. Zasowski, C. Steiner, F. Troesch, and A. Wittneben, *IEEE International Conference on Ultra-Wideband*, ICUWB 2008, Hanover, Germany, Sept. 2008

12. **Ultra Low Power UWB Modem Design: Experimental Verification and Performance Evaluation**
   C. Steiner, H. Luecken, T. Zasowski, F. Troesch, and A. Wittneben, *Union Radio Scientifique Internationale*, URSI 2008, Chicago IL, USA, Aug. 2008

13. **Stick based speckle reduction for real-time processing of OCT images on an FPGA**
   H. Luecken, G. Tech, R. Schwann, G. Kappen, *Acta Polytechnica*, Scientific Journal of Czech Technical University Prague, 2007

14. **Real-time Speckle Reduction in High-Resolution OCT Imaging**
   H. Luecken, G. Tech, R. Schwann, G. Kappen, *Poster 2007*, Prague, Czech Republic, May 2007

15. **Routing Criterion for XPM-Limited Transmission in Transparent Optical Networks**
   S. Herbst, H. Luecken, C. Fürst, S. Merialdo, J.P. Elbers, C. Glingener, *ECOC-IOOC 2003*, Rimini, Italy, Sept. 2003

Bisher erschienene Bände der Reihe

Series in Wireless Communications

ISSN 1611-2970

Alle erschienenen Bücher können unter der angegebenen ISBN-Nummer direkt online
(http://www.logos-verlag.de) oder per Fax (030 - 42 85 10 92) beim Logos Verlag
Berlin bestellt werden.